Maple for Trigonometry

AF616786

RICHARD PARKER

Delmar Publishers

an International Thomson Publishing company ITP®

Albany • Bonn • Boston • Cincinnati • Detroit • London • Madrid
Melbourne • Mexico City • New York • Pacific Grove • Paris • San Francisco
Singapore • Tokyo • Toronto • Washington

NOTICE TO THE READER

Publisher does not warrant or guarantee any of the products described herein or perform any independent analysis in connection with any of the product information contained herein. Publisher does not assume, and expressly disclaims, any obligation to obtain and include information other than that provided to it by the manufacturer. The reader is expressly warned to consider and adopt all safety precautions that might be indicated by the activities described herein and to avoid all potential hazards. By following the instructions contained herein, the reader willingly assumes all risks in connection with such instructions. The publisher makes no representations or warranties of any kind, including but not limited to, the warranties of fitness for particular purpose or merchantability, nor are any such representations implied with respect to the material set forth herein, and the publisher takes no responsibility with respect to such material. The publisher shall not be liable for any special, consequential, or exemplary damages resulting, in whole or in part, from the readers' use of, or reliance upon, this material.

COPYRIGHT © 1997
By Delmar Publishers Inc.
An International Thomson Publishing Company
The ITP logo is a trademark under license

Printed in the United States of America

For more information, contact:

Delmar Publishers Inc.
3 Columbia Circle, Box 15015
Albany, New York 12212-5015

International Thomson Publishing Europe
Berkshire House
168-173 High Holborn
London, WC1V7AA
England

Thomas Nelson Australia
102 Dodds Street
South Melbourne, 3205
Victoria, Australia

Nelson Canada
1120 Birchmont Road
Scarborough, Ontario
Canada, M1K 5G4

Delmar Staff:
Publisher: Robert D. Lynch
Editor: Paul Shepherdson
Production Manager: Larry Main
Art & Design Coordinator: Mary Beth Vought

International Thomson Editores
Campos Eliseos 385, Piso 7
Col Polanco
11560 Mexico D F Mexico

International Thomson Publishing GmbH
Königswinterer Str. 418
53227 Bonn
Germany

International Thomson Publishing Asia
221 Henderson Road
#05-10 Henderson Bldg.
Singapore 0315

International Thomson Publishing Japan
Hirakawacho Kyowa Building, 3F
2-2-1 Hirakawa-cho
Chiyoda-ku, Tokyo 102
Japan

All rights reserved. No part of this work covered by the copyright hereon may be reproduced or used in any form or by any means—graphic, electronic, or mechanical, including photocopying, recording, taping, or information storage and retrieval systems—without the written permission of the publisher.

1 2 3 4 5 6 7 8 9 XXX 02 01 00 99 98 98 97 96

Library of Congress Cataloging-in-Publication Data

Parker, Richard, 1938
Maple for Trigonometry
p. cm.
Includes index
ISBN 0-8273-7409-7
1. Trigonometry- computer-assisted instruction. 2. Maple (Computer file) I. Title
QA 155.8.D4P35 1997
512.9′ 0088 - dc20

Contents

Online Services

Delmar Online
To access a wide variety of Delmar products and services on the World Wide Web, point your browser to:
http://www.delmar.com
or email: info@delmar.com

thomson.com
To access International Thomson Publishing's home site for information on more than 34 publishers and 20,000 products, point your browser to:
http://www.thomson.com
or email: findit@kiosk.thomson.com

Internet

To the Student

A short discussion about the organization of this book will help you get started quickly. The author's objective is to take you through a study of the major results of trigonometry and to have you concentrate on the important details. The study of every topic in mathematics has been made much easier in recent years. The electronic calculator takes care of routine calculations, and you will use Maple® to automate plotting, solving equations, and simplifying algebraic and trigonometric expressions. There is still no substitute for clear thinking. The answer on your calculator is only as accurate as the numbers you put in. Likewise, Maple does what you tell it to do (almost always), but it does not lead you by the hand to the solution. You are in the driver's seat. Consequently, you must develop the confidence to use mathematical tools with authority—and *Maple for Trigonometry* will show you the way.

In this book, a topic is introduced, and then examples are given. You work through a progressive set of exercises. The "paper and pencil" solution to a problem is given first, and then you are shown how to solve the same problem using Maple. You might ask, "If I learn the method of doing the problem by hand, why do I need Maple?" The answer to that question has one main part and some other related ones as well. Most importantly, learning how to do the problem by hand insures that you understand the method: where the theory applies and how the solution is accomplished. But once you have done a few problems by hand and have mastered the concept and its application, you still must go through all the steps for any new problem of the same type that may come along. Maple operates through a computer, and you can make use of the computer's memory. When you translate the problem to a Maple worksheet (which we will define later), you have an active way of solving any similar problem. Just type in the numbers for the new problem and Maple works out the new solution. You don't have to "reinvent the wheel" each time you encounter a new set of numbers. Thus, as you learn each new concept in this book, you will learn how to make up a Maple worksheet for a typical problem and store it on disk. If you need to solve a new problem of a type already seen, all you have to do is to call up the worksheet, change the numbers, and have

Maple recalculate to get the new answer. You will have saved time, and your answer has a much better chance of being right the first time.

Using Maple has other advantages as well. Today, computers are used in all the technological disciplines. The complexity of most technological problems requires millions of computations, and mathematics is no different from any other technology in this regard. Learning to use Maple shows you how to apply computer technology and reap the benefits of having a computer do the tedious number crunching. Plotting a graph is a good example. Imagine what you need to do to plot a graph by hand. You need to decide on the scale for the x and y axes, and then you need to compute a table of 10, 20, or 50 values for x and y. You have to plot these values, one pair at a time, on graph paper. It could take you an hour to do one properly drawn graph. However, Maple can draw a graph in seconds, allowing you to concentrate on interpreting the result. By freeing you from the drudgery of plotting points, Maple lets you use your time more creatively. For example, you can spend time interpreting the graph and making sure you can relate the shape of the graph to the original problem.

There are plenty of "Your Turn" problems in the book, which are designed to give you practice in solving problems. You will be shown particular problem-solving techniques in the "Examples" sections, most of which have a "Your Turn" problem for you to try. The problem sets begin with straightforward variations of the topic at hand and progress to ones that test your understanding at a deeper level. The "Explorations" sections are designed to let you exercise your creativity by asking you to extend the theory or invent some novel way of getting the solution.

The Origins of Trigonometry

Mathematics begins when intelligent beings start to count. Ivars Peterson, in his book, *Islands of Truth: A Mathematical Mystery Cruise,*[1] reports on the work of archaeologist Denise Schmandt-Besserat, who investigated the earliest human use of clay by going from museum to museum and comparing the clay collections of Middle Eastern cultures that thrived between 10,000 and 6,000 B.C. Surprisingly, she discovered a host of little baked clay objects that resemble marbles, disks, and dice. Mathematically, these objects had the geometric shapes of cones, spheres, disks, cylinders, and little pyramids. Obviously, as long as 10,000 years ago, humans were using objects that could be studied trigonometrically. What was the purpose of these small objects, so many of which have survived? Interestingly, other archaeologists had concentrated on the geometric shapes, guessing that the disks were lids for small jars (an ancient bottle cap) and the spheres were marbles, a favorite children's pastime all through the ages (until the advent of the video game). Schmandt-Besserat's great insight was that these objects all belonged together: they shared a common function. After years of investigation, she realized that these were tokens, which were connected with counting and numbers. A great story emerges here. This was the dawn of the agricultural age, when humans began to "settle down" instead of living as free-ranging hunter-gatherers. As communities began to produce food rather than just collecting it, they needed some way of keeping track of this shared bounty. Along with this newfound ability to produce more food, a sharp increase in population occurred. Wandering tribes took root with their vegetables. Human settlements became villages and towns, and rather quickly, some grew to the size of cities. Ancient accountants kept track of the stored bounty by maintaining a "ledger" of baked clay tokens—one token for each item. Differently shaped tokens stood for different groceries: a marble-sized clay sphere stood for a bushel of grain, whereas an egg-shaped token stood for a jar of oil. This system of accounting for agricultural produce

1. Ivars Peterson, *Islands of Truth: A Mathematical Mystery Cruise* (New York: Freeman, 1990).

persisted practically unchanged for almost 4,000 years. In comparison, our modern era has been in existence for about 400 years, just one tenth as long!

These agrarian kingdoms began to construct monuments and great temples, beginning in the middle of the fourth millennium B.C. The construction of these great public structures required the organization of many workers and of huge amounts of materials. The token system was extended to cover these goods and services. Human inventiveness had modified a system for keeping track of grain and other foodstuffs so that the demands of a more diverse society could be met. Here the power of even the most elementary mathematics can be seen. These tangible counters are the forerunners of the modern ledger, where all sorts of business accounts are kept. Once people realized that they could make a correspondence between two very different things, such as a bushel of grain and the tiny token representing the grain, the seed for the development of elaborate mathematics had been planted. This mathematics was rooted in very practical things, such as the need to record the amount of the harvest and to keep track of building materials. It would be many years before someone would study mathematics for its own sake, independent of any particular application.

The material we will study in trigonometry is almost exclusively of the practical variety. An entirely practical approach is, perhaps, difficult to accomplish because the foundation of any mathematical study involves definitions. You must become familiar with mathematical terminology, and you need to do a number of representative problems to master the concepts. All the same, trigonometry has application in almost every field of technology, science and engineering. Surprisingly, it also pops up in business studies and medicine. Trigonometry is at the foundation of mathematics so it should not surprise you that the world is teeming with trigonometric applications. Of course, the first of these applications came from surveying the land.

As people settled the land, a system of measuring it became necessary. At first, we may imagine that small groups would begin tilling the soil in some favorable area. As the community prospered, it grew. Eventually, it would be so big that a system of measuring the land and recording those measurements would be required. The science of trigonometry would be born from these attempts to lay out and measure the fields belonging to various groups.

You may be tempted to conclude that the study of trigonometry is only useful if you intend to take up surveying as a career, but this is not the case. Those mathematicians who studied the subject for its own sake have made us realize that subjects as widely separated as automotive mechanics, electronics, and even quantum theory require results from trigonometry for their understanding. It is the wonder of mathematics that it can be applied to so many different tasks.

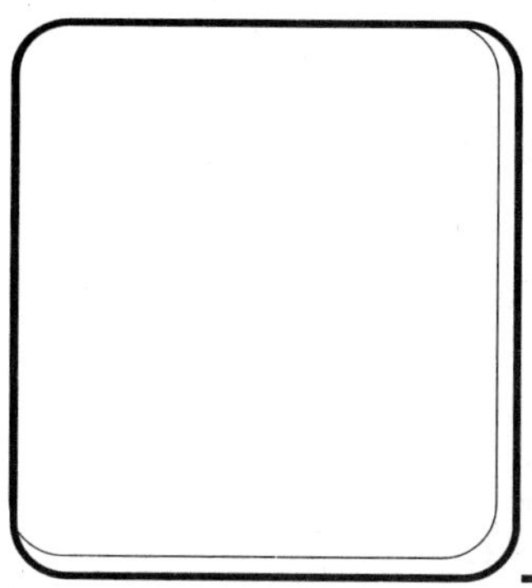

Prelude to Maple

Objectives for This Chapter

This book is about using Maple to enhance your mathematical expertise. Each chapter shows you how to use a handful of Maple commands that bear upon the chapter topic. But before you start using Maple on these topics, you need to understand something about its use. That is the purpose of this introduction.

Maple, like mathematics, is a big undertaking. The best way to tackle such a large project is to start small and do a little more each day. That way, you move gradually ahead. You'll begin the process by getting an introduction to Maple here. It is definitely not meant to be a complete tutorial, it is only meant to get you started. The whole book is the tutorial! When you have completed it, you will be able to apply mathematics to a wide range of trigonometric topics and you will have the ability to verify the results of your calculations.

The specific objectives for this section are listed below.

1. Learn the purpose of a computer algebra system, such as Maple
2. Know how to issue a command to Maple
3. Know how to avoid some common pitfalls when issuing commands
4. Confirm your knowledge by taking a quiz

What Is Maple?

Maple is a kind of super-calculator called a *computer algebra system*. If you are using a computer with a graphical user interface, like the Macintosh®, Windows®, or an X-terminal, you will be using Maple's worksheet interface. The worksheet provides an integrated environment in which you can perform mathematical operations, use a simple word processor to explain your calculations, and include Maple plots, as well as allowing you to store formulas, equations, and other symbolic mathematical structures for use later on. Of course, you can do strictly numerical calculations as well. These may sometimes be a little more cumbersome than on your pocket calculator, but there are advantages. You can compute to many more decimal places than on your calculator, and you can save your results on disk or print them as part of a written report. This ability to store your results and use them again later on, perhaps with modifications, is a great advantage once you learn how it's done. Most important, there are techniques for verifying that your calculations are correct. This indispensable step is not nearly as easy on a calculator as it is with Maple.

Computer algebra systems are important tools for doing mathematics. They can help you use mathematics with confidence. The most important step in any problem is checking the answer. Does your answer really satisfy the problem you started out to solve? Using Maple, it is very easy to substitute the solution back into the original equations and see that everything checks, whereas doing this step by hand, even if you use a calculator, sometimes requires so many manipulations that the probability of a slipup becomes depressingly high. There is a temptation to say, "Oh, well, I must have gotten the right answer, so I'll go with it." However, later, after you leave school, there will be a lot more riding on the answer. You will want your calculations to be correct. That means you will want to use computational tools that permit you to record your work and allow you to verify your conclusions.

Mathematics has been under more or less continuous development for 2,500 years. There is an astonishing amount of accumulated knowledge, which the designers of Maple have tried to incorporate. Even so, you can use Maple effectively once you have learned only a handful of commands. This tutorial is meant to get you going. Further examples will be found throughout the book and you can use Maple's own help function to browse through the system's capabilities. Many people begin with the sections on plotting, especially the three-dimensional plot and animation packages. The best advice is—go ahead and enjoy!

Prelude: Before You Begin Maple

You are familiar with the pocket calculator, and you may be familiar with programming languages such as PASCAL, C, or C++. All these use approximations for real numbers. Typically, the accuracy of their numerical engines is about 10 or 12 significant figures. For instance, the difference,

$$12345678901234567789 - 12345678901234567788$$

yields zero on my calculator, whereas the answer clearly is 1, as you can see by inspecting the two numbers. This occurs because all computations on these calculators are only approximate. Most calculators will refuse to accept 19-digit numbers, and those that do simply throw away anything after the 12th digit or so.

In contrast, Maple has been designed to treat numbers as *exact*. This will lead to confusion in first-time users, to whom it may seem that Maple is not working out an answer. As an example, take the mind-bogglingly easy problem of working out 1/3. If you type this into Maple as follows,

> 1/3;

$$\frac{1}{3}$$

you see that you get the same thing, just printed a little nicer as a fraction, the way you would see it printed in a textbook. If you think about this result for a little, you realize that one-third is the *exact* answer, while 0.33333333 is approximate, because in reality, in the decimal representation of one-third, the 3s go on forever.

Some Frequently Asked Questions

1. *Maple is great at algebra, but how do I work with decimal numbers?* One question that very quickly arises is, "How do you get the decimal equivalent to a number that Maple has specified exactly, such as 1/3, or sin(30)?" There is a way! In fact, there are many ways, and you will learn about them in the book. This means that you can use Maple as a calculator as well as doing symbolic manipulations with it. Don't be concerned that Maple tends to write results as fractions rather than decimals. It's easy to convert any of Maple's numbers to decimal form. You will be given many examples in this text.

2. *How do I issue commands to Maple?* Maple is an integrated package. You can use it as a simple word processor as well as a calculator. Therefore, the next question that needs answering is, "How do I tell Maple to compute something?" First, you ensure that you are in Maple input mode—that is, you see the Maple prompt (>) at the beginning of the line—then, you type in the expression you want to compute, and you remember to add a *semicolon* (;) at the end. The semicolon tells Maple to work on what you have typed in. If you don't put in the semicolon, Maple will do nothing! It will wait until you have typed in a semicolon before it attempts to work on what you have typed.

3. *What do I do when things seem to be going wrong?* The third thing to remember is this: if you have been progressing through the calculations yet Maple seems to be giving you the wrong answers or is doing nothing, the best thing to do is to start over. Quit Maple and restart it. Maple tries to remember everything you have done in order

to save time if you choose to repeat a calculation. But if you have typed something that has confused Maple, it will remember that too! This can cause all sorts of problems for new users. The best approach is to start over again rather than digging yourself into a deeper hole.

4. *I hate typing. Won't this slow me down?* The purpose of this book is to show you how to get reliable answers. An important part of the process is to keep typing to an absolute minimum. The book will show you how to use the copy and paste functions instead of retyping mathematical input. This eliminates the chance of error when you need to use the same expression again on a new input line.

5. The fifth and last general point concerns *naming objects in Maple*. Let's say you have typed in a quadratic equation. Later on, you want to use that equation again. Do you have to type it in again? Not if you gave it a name the first time. Once a complicated object has been assigned a name, all you have to do is use that name and Maple will replace the name with the object. Get in the habit of naming every expression you type into Maple. Here is an example of giving something a name. We will name the quadratic equation $6x^2 + 7x - 20 = 0$ by calling it Moe. Here's how:

> Moe := 6*x^2+7*x-20 = 0;

$$Moe := 6x^2 + 7x - 20 = 0$$

You name an object by using the *colon equal* assignment operator (:=). There must *not* be a space between the colon (:) and the equal sign (=). Think of this operator as one word. Furthermore, notice that because you have given a name to an equation, the equation contains an equal sign. The *colon equal* is an assignment operator, whereas the *equal sign* has its ordinary meaning in mathematics. The name *Moe* now stands for this equation. You can verify this by typing in this name on a Maple input line:

> Moe;

$$6x^2 + 7x - 20 = 0$$

Maple looks up the name *Moe* in its memory and sees that the name stands for an equation. It prints out the equation as the "value" of the name.

If you make a typing error and write *moe* instead of *Moe*, Maple responds as follows

> moe;

$$moe$$

because the name *moe* has not been defined. Be careful with all your typing, as Maple is

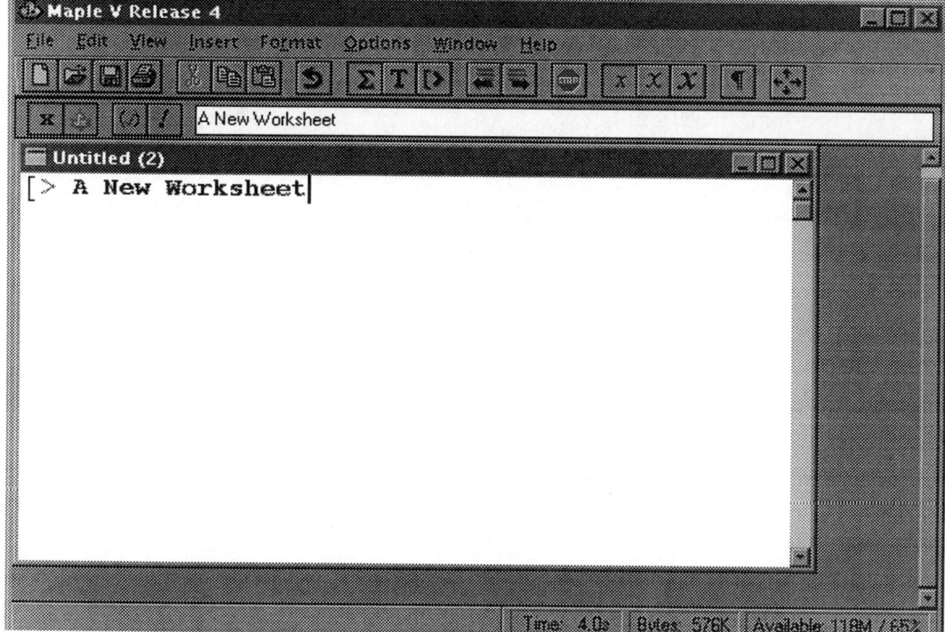

Figure A.1 The Worksheet Interface in Release 4

case sensitive. Clearly, naming objects makes it easy to refer to them again and again, so get in the habit of giving a name to any expression you might ever use again.[1]

The Worksheet

You' ll probably start Maple by double clicking on the Maple icon. When you start Maple, the screen will look like Figure A.1 if you are using Release 4.

---

[1] You can reexecute a Maple command line by placing the cursor on a previously executed line and editing it. In particular, you can "wrap" _____ another command "around" the command. Say your command was

>2*Pi*33.2;

You realize when you see the output of this command that you need to wrap an *evalf* around the command. Place the cursor before the *2* and type *evalf(*. Then move to the end of the command and close the brackets. You now have

>evalf(2*Pi*33.2);

Press Return, and the new command replaces the old. You can "build up" commands in this way, which allows you to check that intermediate results are what you want. (The *evalf* function is introduced in detail later in this Prelude.)

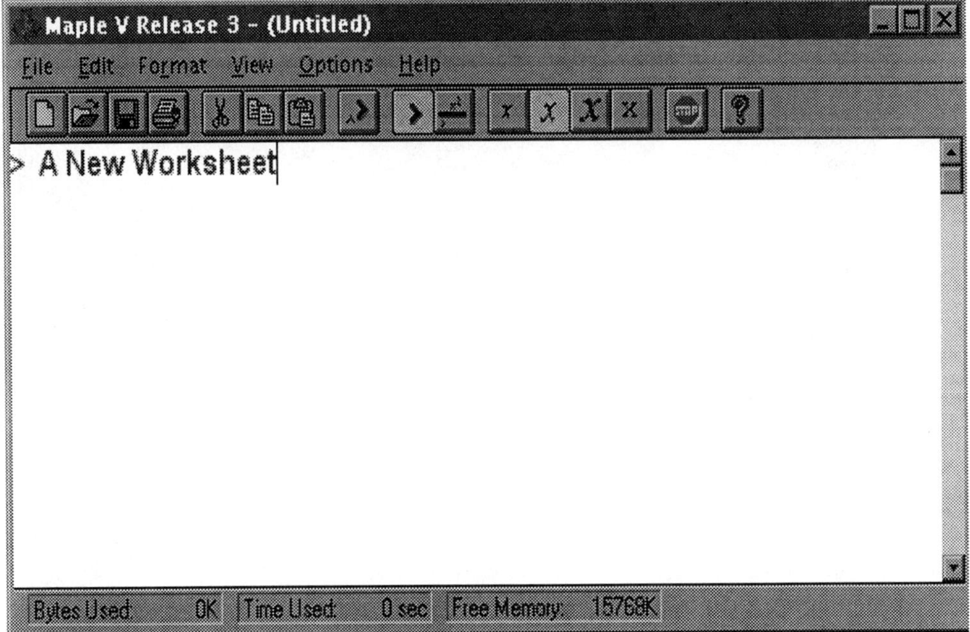

Figure A.2 The Worksheet Interface in Release 3

If you are using Release 3, the screen will look like Figure A.2.

We won't explain the function of all the buttons here. You'll become familiar with them as you work with Maple. If you have Release 4, there is one button that you should use from the beginning. It is the **x** button on the *context bar*. You have to type mathematical statements on a single line using the keyboard, but clicking on this button allows you to see what you have typed in true mathematical form. You can check your typing and make sure that you have input the correct expression by looking at the typeset result. Release 3 does not have this capability. If you are using Release 3, it is a good idea to type the expression you are going to work with on an input line (terminated by a semicolon and a return) *before* you apply any Maple operations to it. This technique permits you to see the typeset form of your input so you can check it for accuracy. Only when you are convinced that you are working with the right expression should you continue performing other operations on it. After all, if you begin wrong . . . !

Your First Maple Commands

Maple is a calculator, which you can use to do arithmetic. Type these expressions into Maple and observe the results.

Addition

> **3+4;**

```
7
```

> **7+11;**

```
18
```

> **12345678901234567890 +12345678901234567890;**

```
24691357802469135780
```

Be careful when typing in this last command. Notice that the numbers are simply the numerals 1 to 10 typed in twice. In Maple, the result is computed to full accuracy. Pocket calculators don't give 20-digit accuracy.

Let's Practice. Use Maple to add the first 10 odd numbers. Here's the command

> **1+3+5+7+9+11+13+15+17+19;** Answer:____________________

Now, find the sum of the first 11 odd numbers. *You don't need to retype!* Copy the previous command and paste it on a new input line. Place the insertion point between the *19* and the semicolon and type *+ 21*. By copying and pasting, you have used your previous work to eliminate almost all the typing for your new command.

> **1+3+5+7+9+11+13+15+17+19+21;** Answer:____________________

Subtraction

> **1234567890123456789 - 1234567890123456788;**

```
1
```

Here, the numbers differ by 1 in the units column; otherwise, the digits are the same. The result must be 1, since for any two numbers where one is one more than the other, $n + 1 - n = 1$.

Let's Practice. Subtract these numbers: $1 - (2 - 3) - (4 - 5) - (6 - 7)$. Check your result by mental arithmetic.

Answer: ____

Multiplication

> **3*4;**

12

> **25*34;**

850

> **1.2*7.4;**

8.88

This last example shows you that if you put a decimal point in the input numbers, Maple does give you a "real," or "floating," point (decimal) answer. More examples will follow under "Subtraction."

> **0.0000003 * 7;**

$.21\ 10^{-5}$

This is an example of how Maple writes an answer using powers of ten notation. Maple usually writes large or small decimal numbers as in this example, but it may use the float notation if the numbers get too small.

> **1234567890 * 1234567890;**

1524157875019052100

Let's Practice. Multiply the first six integers together. Remember to put in the multiplication sign (*).

Answer:____

Division

> **1/3;**

$$\frac{1}{3}$$

> **1/3.;**

.3333333333

Do you see the difference in these commands? The first uses whole numbers, and the second uses a "real" number. There is a decimal point after the 3, so Maple interprets this number as 3.0, a decimal number. Once there is a decimal number in the expression, Maple switches to floating-point format rather than treating each number as exact. It gives its answer in decimal form, using ten digits as the default.

For example:

> **355/113; 355. / 113;**

$$\frac{355}{113}$$

```
.3141592920
```

The above fraction is an approximation to the number π, which is the ratio of the circumference of a circle to its diameter.

Let's Practice. Do this division problem in Maple: 1/2/3/4. If you are using Release 4, click on the *X* button at the far left of the context bar. Interpret what you see.

Answer: ______________________________

Where are the implied brackets? Are they (a) ((1/2)/3)/4, or (b) 1/((2/3)/4)?

Answer: ______________________________

Exponentiation

To raise one number to a power, use the "up arrow" (SHIFT+6) on the keyboard (or use **).

> **3^2;**

```
9
```

> **5^2.2;**

```
34.49324154
```

> **25^25;**

```
88817841970012523233890533447265625
```

These last two examples show that you can use a decimal number as the exponent, and that you can raise numbers to very large powers if you need to.

Let's Practice. Which is bigger, 22^{25} or 25^{24}? Answer: ______________

Using *evalf* to Convert an Answer to Decimal Form

You know from some of the examples above that you can get Maple to work in floating point form by putting one decimal number in the expression. Just one decimal number will cause Maple to switch to decimal format. In that case, the answer will be approximate, but it will be in the familiar decimal notation. Another way to convert numbers is to use the *evalf* function. You will be using functional notation for many Maple operations.

The general format of a function call in Maple is shown by the *evalf* function. As an example, take the number π. Recall that this number is the ratio between the circumference and the diameter of a circle. If we input this number in Maple, we get:

> **Pi;**

$$\pi$$

Maple understands this input and writes the standard symbol for Pi. If we type:

> **pi;**

$$\pi$$

you might think that Maple treats this *pi* the same, but this is not true! It is important to note that Maple does make a distinction between upper-case letters and lower-case ones. This last result is "the Greek letter *pi*," not the ratio!

To find the decimal approximation for Pi, we use the *evalf* function:

> **evalf(Pi); evalf(pi);**

```
3.141592654
```

$$\pi$$

Pi has a value of approximately 3.14, whereas the Greek letter *pi* has no decimal value.

The factorial

Maple knows how to find the factorial of a whole number:

> **3!;**

```
6
```

> **7!;**

```
5040
```

> **45!;**

```
119622220865480194561963161495657715064383733760000000000
```

The factorial function certainly grows quickly!

Integer Factorization: The *ifactor* Function

Maple contains many functions that help us work with whole numbers, among which is the *ifactor* function.

> **ifactor(5670);**

$$(2)\,(3)^4\,(5)\,(7)$$

It gives all the prime factors of a number.

Let's Practice. What is the decimal equivalent of Pi/3? Can you get the answer by typing:

>**Pi/3.0?**

Answer: ____

Remaindering

This is something that is not easy to do on a calculator. Say you want to find the remainder when 65537 is divided by 13.

> **irem(65537, 13);**

4

The remainder is 4. One way to prove this is to take the quotient, using the function *iquo*:

> **iquo(65537, 13);**

5041

You can check the result by reversing the process:

> **5041*13 + 4;**

65537

You can take the remainder of some very large numbers, a thing you cannot do easily using a calculator.

> **irem(45!, 53);**

32

Exercise: What is the remainder when you divide 45! by 11, by 17, by 44?

Maple's Help Facility. There are many other functions that you can use to investigate integers. This is a good place to introduce Maple's *help* facility. The syntax is shown below. Type a question mark and the name of the thing you want help for. You don't need to use a semicolon in this case. You will get a separate help window. If you don't get the help topic you want the first time, look at the "See Also" section of the Help screen. It may give you a good hint on where to look next. Try the command *?integer* to see the help window for a list of integer functions (see Table A.1).

> **?integer**

Table A.1

Maple Help Window for ?integer

Help For: Integers

Description:

An expression is of type integer if it is an (optionally signed) sequence of one or more digits of arbitrary length. The length limit of an integer is system dependent but generally much larger than users will encounter—typically greater than 500,000 decimal digits.

In addition to arithmetic operators, other basic functions of integers are

abs sign min max factorial

irem iquo modp mods mod

isqrt iroot isprime ifactor ifactors

igcd ilcm igcdex iratrecon rand

There are also many special functions for integers in the numtheory and combinat packages including the binomial coefficients, Fibonacci numbers, Stirling numbers, Jacobi symbol, Euler' s totient function, etc.

See Also: type, numeric, constant, numtheory, combinat

Prime Numbers and Factoring

You may recognize some of these functions, while others may be a mystery to you. You can find out about each one by using the help function again. One of the integer functions is *isprime*. If you ask for help on this topic, you will see that it gives a *true* or *false* answer, depending on whether the number you give it is prime (has no factors) or not:

> **?isprime**

> **isprime(65537);**

```
true
```

The fact that 65537 is prime is not too hard for you to figure out for yourself if you take a few minutes, but what about a really big number?

> **isprime(65537^6 + 1);**

```
false
```

How can you show that this number has factors? You would multiply it out, add 1, and then factor it:

> **(65537^6 + 1);**

```
792354163458888160381945774l0
```

> **ifactor(65537^6 + 1);**

```
(2) (5) (18447869995091361793) (49477) (8681)
```

By looking at the expansion of the number, you see it is even. Therefore, it must have a factor of 2. Its digits end in 10, so it must also be divisible by 5. Don't make Maple a substitute for your own analysis of a problem. Use Maple to assist you in doing calculations, but first think about the problem carefully to see if you can grasp the answer directly, or at least make some significant simplifications.

Let's Practice. Which of these numbers is a prime? $2^2 + 1$, $2^3 + 1$, $2^4 + 1$, $2^5 + 1$, $2^6 + 1$, $2^7 + 1$, $2^8 + 1$

Answer: ______________________________

Real Numbers

You have encountered the real number line before. Besides integers and fractions, the real number line contains the so-called "irrational" numbers like the square root of 2. In Maple, you can obtain a square root by using the *sqrt* function. Let's take the square root of 65,536.

> **sqrt(65536);**

```
256
```

The original number, 65,536, is a perfect square, the square of 256. What if we attempt to take the square root of 65,537?

> **sqrt(65537);**

$$\sqrt{65537}$$

Maple simply writes out the number again, using textbook mathematical notation. Note the difference between the *sqrt* function and the *isqrt* function:

> **isqrt(65537);**

```
256
```

The *isqrt* function gives you the nearest whole number that will square. But what if you want the "answer" to the square root of 65,537? Remember what question you are asking! The square root of 65,537 is simply what Maple gave you: 65,537 under the square root sign. This is the completely precise answer. Anything else is an approximation! Sometimes, however, it is the numerical approximation that you want, and in this case, remember that there are two ways to get it: either put in the number as a decimal number:

> sqrt(65537.);

```
256.0019531
```

or use the *evalf* function:

> evalf (sqrt(65537));

```
256.0019531
```

Notice that you can "nest" functions (put one function inside another) in Maple. The result of *sqrt* (65,537) has a value that is given to the *evalf* function. You always read such nested functions from the inside out. This is no different from other "functions of functions" that you have encountered earlier in mathematics, such as $f(g(x))$, which means that the function f is evaluated at the point $g(x)$.

Let's Practice. Find the decimal approximations to $\sqrt{3}$, $\sqrt{5}$, $\frac{1+\sqrt{5}}{2}$:

Answer:____

Assignments: Examples of Names in Maple

To use something like the square root of 65,537 repeatedly, you can give it a name. The name you give it will usually be something fairly short, but you must avoid using names that Maple itself uses. For instance, it is a bad idea to use "sqrt" as a name, because Maple uses it for the square root function. Instead use the letters *a, b, c*, and so forth. You can use capital letters as well, but don't use *D* or *I*. Maple uses these as names itself. You will see how to assign a name to an expression in the next section.

The Circumference of a Circle of Radius R. The formula for the circumference is $C = Pi * D$, where C stands for the circumference and D stands for the diameter. We could write, on a Maple input line,

> C = Pi *D;

$$C = \pi D$$

but we have broken the rule of not using capital D in an expression. Also, we have typed in an equation, *but we have not named anything.* To name $Pi * d$ (we are now using a legal variable for the diameter) as the circumference, we would type in:

> **C := Pi * d;**

$$C := \pi d$$

Notice the difference carefully. We have used the *colon-equals* sign to assign the value *Pi * d* to the name *C*. Read this last sentence repeatedly until the meaning becomes locked firmly in your mind. From now on, *C* is exactly the same as *Pi * d*. To see that this is the case, type *C* on a Maple input line:

> **C;**

$$\pi d$$

Maple responds by outputting *Pi * d*, because that is what *C* is "equal to." More accurately, *C* is a name for the expression *Pi * d*. Compare this with the equation for a straight line,

> **y = m*x + b;**

$$y = mx + b$$

> **y;**

$$y$$

Here, we have typed in an equation without naming it. In this case, *y* has not changed its value; it is still *y*, and is part of an *equation*, whereas *C* is a *name*. It has been *assigned* a value, *Pi * d*, by using the colon-equals assignment statement. Note as well that you can give an equation a name. Say we want to give the equation for a straight line a name:

> **a := y = m*x + b ;**

$$a := y = mx + b$$

We have used both the colon-equals and the equal signs in the same line. If we now ask Maple what *a* is, we get:

> **a;**

$$y = mx + b$$

The variable *a* is just the name for the equation, $y = m\,x + b$. To reuse the equation, we can either type out the whole equation or simply use the name.

The Substitute Command

If circle is defined as we have just done, how do we calculate a circumference given a particular diameter? We use the *subs* (substitute) command. In words, we substitute the given value of the diameter into the formula and evaluate. In Maple, we use the substitute command, *subs*.

Exercise: Calculate the circumference of a circle if the diameter is 6.

Solution:

> subs(d = 6. , C);

$$6.\pi$$

In this case, putting in a decimal value was not enough to get a numerical answer. Instead, we must use *evalf.*

> evalf (subs(d = 6, C));

```
18.84955592
```

The answer is 18.85 units.

You will be tempted to think that the calculator method is a lot simpler for this problem, and of course you are right. But the method shown here is important in more complex situations, and you will be using it to solve many problems that are not possible to solve on the calculator.

Let's Practice. Given the polynomial $x^3 + 3.6111\,x^2 - 13.30893x + 1.153922$, find its value when $x = 0 - 10.0, -5.9, 0, +0.889, +2.2, 5, 10$. Define the expression, then use the *subs* command.

Answer:____

Algebra

Maple has an extensive set of commands for algebra. This book concentrates on results, not manipulations. One reason for this approach is that most people need to work within a given mathematical framework where the basic formulas are already given in a particular form. I believe that it is more important to be able to use a given expression (or equation) to accomplish goals, which are usually (a) computation, (b) solving equations, (c) plotting, or a combination of all three. Maple is an ideal assistant here. There are other software packages, like Theorist,® that give you control over each algebraic step in a transformation. Such packages are useful if you want to transform expressions from one form to an equivalent one, although Maple can be persuaded to do step-by-step calculations by experienced users.

The computation may be symbolic or numeric. You may need to find the solution to a quadratic equation, or you may want to picture the behavior of a function over a range of values. This means computing the value of a function at a point, or, if a set of points is required, a graph. You may want to get an overview of the function by plotting it and then computing the value of the function at a few points to high accuracy. A plot is the best way of appreciating the behavior of a function over a set of points.

The starting place for all these operations is the function (or expression). Pay extreme attention to brackets when you type in your formulas. Note the difference between what you want and what you get in the following examples.

Grouping—Brackets. In electronics, a quantity called the (magnitude of the) capacitive reactance is found using the formula $\frac{1}{2\pi fC}$. Enter this formula into Maple.

> **1/2*Pi*f*C;** *(Wrong!)*

$$\frac{1}{2}\pi f C$$

Paying proper attention to brackets, we type

> **1/(2*Pi*f*C);**

$$\frac{1}{2}\frac{1}{\pi f C}$$

The rule to remember is that the division operator (/) applies only to the next term and not to the rest of the multiplicative terms unless you enclose all of them in brackets.

Another need for brackets occurs when using fractional powers. For example, find the cube root of 27.

> **27^1/3;** (*Wrong!*)

$$9$$

This is an appealing answer, which might lure you into accepting it! However, what you really wanted is:

> **27^(1/3);**

$$27^{1/3}$$

Now you see that you have typed the expression in properly, because Maple shows you how it interpreted your input. Unfortunately, it merely wrote it out again without simplifying it. The solution is to use the *simplify* command.

> **simplify(27^(1/3));**

$$3$$

Maple needs some encouragement in getting this result. The reason? There are three possible answers to a problem involving cube roots, so Maple preferred to keep the expression in its original form until it got some hint on how to proceed. The principal root was then found by using the *simplify* command.

Spaces—Where Not to Put Them. Give the name *eq1* to the equation $s = vt$. We will put a space between the colon and the equal sign. Look carefully!

> s : = v*t; (*Wrong!*)

`Syntax error, `=` unexpected` [Release 4]

`syntax error` [Release 3]
```
> s : = v*t;
        ^
```

Maple lets you know that a typing error has occurred. It even gives you some clue about where it has occurred. In Release 3, it puts an upward arrow (^) underneath the place on the input line where it detected the error. In Release 4, it puts the blinking cursor on the syntax error.

On the other hand, Maple does allow spaces as long as the meaning is preserved. You just can't put a space in the middle of a word (the colon-equal sign is a word).

> sqrt (x+ y/ z); (Spaces are allowed here because they are between the words.)

$$\sqrt{x + \frac{y}{z}}$$

There Is No Implied Multiplication in Maple. Many calculators allow you to omit the multiplication sign in certain cases, and it is usually omitted in math textbooks. This is a convention that works only if variable names are restricted to one letter. Maple allows variable names to be very long, so you must *always* put in the multiplication sign. If you had typed in "> s := vt;" for the previous formula, Maple would treat it as a legal expression.

> s := vt; (Forgot the multiplication sign!)

$$s := vt$$

Everything looks just fine, but you could be in for a nasty surprise. There is only one variable in the above formula and its name is *vt*. Rather than *v* multiplying *t*, there is the single variable whose name is *vt*. To ensure that you understand the difference, examine the two commands

> s1 := vt; s2 := v*t;

$$s1 := vt$$
$$s2 := v\ t$$

There is a slight difference in the output. In *s*2, there is a small space between the *v* and the *t*, indicating an implied multiplication sign. Let us try and substitute $v = 5$ and $t = 3$ in these expressions.

> **subs(v=5, t=3, s1); subs(v=5, t=3, s2);**

```
vt
15
```

Since there is no *v* (or *t*) in the first expression (s1), Maple leaves it unchanged. Maple does find a *v* and a *t* in expression s2; thus, the substitution works and Maple multiplies the two to get 15.

Watch the Caps-Lock Key. Maple is case sensitive, which means that you cannot type SQRT when you want to take the square root. Note the difference:

> **SQRT(16), sqrt(16);**

```
SQRT(16),4
```

The second command works, producing an answer of 4, but the first command is interpreted by Maple as an undefined function, SQRT. Since Maple does not know the definition of SQRT, it makes no attempt to evaluate it, but the command does not produce a syntax error. After all, you might define SQRT later on in the worksheet.

Terminate Commands with a Semicolon. A typical situation is represented here. Imagine that you typed a command and pressed Return without terminating the command with a semicolon. Release 4 will warn you to put in the semicolon. If you go back and put it in, everything will be fine.

In Release 3, nothing at all will happen. You look at the command and realize that you forgot to put the semicolon at the end. You go back to the command, type in the semicolon, and press Return. Maple indicates a syntax error. You examine the command and can't see anything wrong. The temptation is to make changes to the command, which makes things go from bad to worse.

> **3 + 4**

>

After getting this far, you realize that you forgot the semicolon. You move the cursor back, put in the semicolon, and press Return.

> **3+4;**

```
syntax error:
    3+4;
     ^
```

There is nothing wrong with the command or its syntax now! The error occurred because Maple remembered the first *3 + 4* that you typed in. When you went back and added the semicolon, the command became *3 + 4 3 + 4;* because Maple reads the line over again when you press Return. There is a syntax error in the command now, caused by the space between the *4* and the *3*. This is no longer a legal mathematical statement. To avoid the problem, when you notice that you have forgotten a semicolon, type it in on the next line (which will otherwise be blank) like this:

> 3 + 4

>;

7

Trigonometry

The most important thing to remember when using Maple to do problems in trigonometry is that Maple uses radian measure, not degrees. Consequently, you must convert any angle into radians before using it in a Maple calculation. To convert from degrees to radians, multiply the angle in degrees by Pi and divide by 180.

Example: Convert 30 degrees to radians.

> evalf(30*Pi/180);

```
.5235987758
```

Thirty degrees is equivalent to 0.5236 radians.

You can also look up values for sin, cos, tan, and other trigonometric functions:

Example: Find the sine of 30 degrees.

> sin(30);

`sin(30)` *(Wrong!)*

You must convert to radians and use *evalf* if you want a decimal answer:

> evalf(sin(30*Pi / 180));

```
.5000000000
```

If you want the exact answer, simply type in the conversion factor between radians and degrees: Maple will reduce this to sin(Pi/6) and display the exact result.

> sin(30*Pi / 180);

$$\frac{1}{2}$$

Using exact numbers like Pi/6, Maple has given us the exact answer, 1/2, correct to "infinity" decimal places. When Maple gives an answer like .500000000, we cannot be sure whether the zeros go on forever or whether some other digits will show up later on in the expansion. However, when Maple gives the answer 1/2, we know it is exactly 1/2.

You will learn in this book that there are two units of measure for angles. In high school, you probably learned how to measure angles in *degrees*, so by now, measuring angles in degrees must be quite familiar to you. You may have encountered the other common unit, the *radian*. Radian measure is so important in advanced mathematics that most engineering texts freely mix radians and degrees, assuming that the reader will be able to determine from the context which unit is being used. However, some books exclusively use radian measure, and Maple does also. Since we will be using degrees frequently, we need to teach Maple to use these units as well, and here is a good place to start.

You' ll see how to convert from radians to degrees and vice versa in Chapter 1. We will discuss here how to look up the basic trigonometric functions when an angle is given in degrees. If you want to use degrees as a unit of measure while you are using Maple, a simple method is to open a new Maple worksheet and issue these commands.

```
> Sin := z -> sin(z*Pi/180);

> Cos := z -> cos(z*Pi/180);

> Tan := z -> tan(z*Pi/180);
```

Save the worksheet and give it a name, such as *trig*. Whenever you want to do trigonometry with Maple, open this worksheet and execute the three commands. You may want to put in some more commands from Chapter 1 to help you convert from degrees to radians and back. An important distinction is in the name *Sin* (rather than *sin*). Case-sensitive Maple considers these two names to be totally different entities. It reserves the name *sin* (all lower-case letters) for the trigonometric function called the *sine*. To use the *sin* function, you must write the angle in radians. However, Maple does not define a *Sin* function, so you are free to define your own, as we have done in the previous commands.

To use either command, you write the statement just as you would in a mathematical essay. Here are a few examples. Take it as given that the sine of 30° is 1/2. This result follows from the definition of the sine function applied to an equilateral triangle, one of whose angles has been bisected to create the 30° angle. If you want to use Maple to look up the sine of 30°, you would issue the command:

```
> Sin(30);
```

$$\frac{1}{2}$$

This is the expected result. Note what happens if you use the lower-case *sin*:

> **sin(30);**

sin(30)

Mysteriously, Maple simple repeats the input line. You might decide that you should find the decimal approximation to *sin(30),* so you issue the command:

> **evalf(sin(30));**

-.9880316241

After you have made a slip like this a few times, you will realize that > *sin(30);* is the command for looking up the sine of 30 *radians* (approximately equal to 1,719 degrees), not 30 degrees. It's not surprising that the sine of 30 radians is a negative number. After all, 1,719 degrees is nearly 4¾ revolutions. The terminal side of this angle points nearly straight down, where the sine would be –1.

With these commands you have taught Maple how to work in degrees as well as radians. You could include a few more commands, such as:

> **Csc := z -> csc(z*Pi/180);** (The cofunction definitions)

> **Sec := z -> sec(z*Pi/180);**

> **Cot := z -> cot(z*Pi/180);**

> **Arcsin := z -> arcsin(z)*180/Pi;** (Definitions for the inverse trigonometric functions)

> **Arccos := z -> arccos(z)*Pi/180;**

> **Arctan := z -> arctan(z)*Pi/180;**

These definitions permit you to work with degrees in Maple as easily as with radians. Just make sure you use the correct function—the one that corresponds to the units in your problem!

Plotting

To plot a graph in Maple, you use the *plot* command. First you type in *plot (*. Then add the expression to be plotted (not the equation, just the right-hand side of the equation!), then a comma, then the range for the *x* axis. You finish off the command with a closing bracket and a semicolon. Here is an example:

To plot *sin(x)* for *x* ranging from –2*Pi to 2*Pi, type in the following (see also Figure A.3):

> **plot (sin(x), x = -2*Pi .. 2*Pi);**

Note especially how you specify the range for the *x* axis. You type *x*=, then give the left-hand (lower) starting place for the range, type in two dots (periods), and specify the right-hand (upper) boundary for the range.

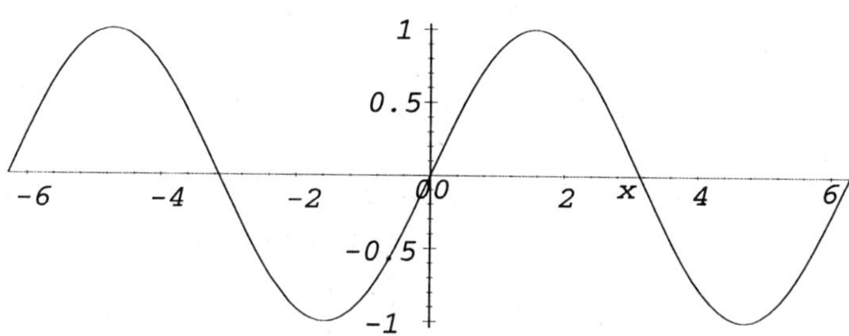

Figure A.3 Maple Result: The Graph of a Sine Wave

This brief introduction to Maple will enable you to get started. You will learn more commands as you work through the book.

Dos and Don'ts When Using Maple

Do end every Maple command with a semicolon.

Do start Maple over again if you get confusing results or no results at all. Maple tends to remember everything, even your mistakes!

Don't use double quotes to refer to the last result, assign names to Maple statements instead, or reuse the command by wrapping (nesting) commands around it.

Don't use:

> a := (1/x + 1/y); normal(");[2]

$$a := \frac{1}{x} + \frac{1}{y}$$

$$\frac{y + x}{xy}$$

Instead, you can use either one of the following two commands.

---

[2] The *normal* command finds a common denominator and writes the expression in this form.

> normal(a), normal((1/x + 1/y));

$$\frac{y+x}{xy}, \frac{y+x}{xy}$$

The second of these commands is an illustration of "wrapping" a command around a previous command. On the other hand, you can't wrap a command around an assignment.

> normal(a := (1/x + 1/y));

```
syntax error:
normal(a := (1/x + 1/y));
           ^
```

Do use the *copy* and *paste* functions to edit Maple commands. Here's an example of simplifying an expression and evaluating a numerical result. The expression, $\frac{1-e^x}{1+e^{-x}}$ + $\frac{1+e^x}{1-e^{-x}}$ is to be simplified and then evaluated at $x = 0.5$.

1. Type the expression into Maple. Use parentheses to maintain the grouping in the original expression:

   > (1 -exp(x)) / (1 + exp(-x)) + (1 + exp(x)) / (1 - exp(-x));

   $$\frac{1-e^x}{1+e^{(-x)}} + \frac{1+e^x}{1-e^{(-x)}}$$

2. Check Maple's output and ensure that it is the same as the original problem. In Release 4, you can use the *X* button on the Context bar to see the typeset form of your input.

3. Now that you have ensured that the input is correct, highlight the Maple input line and copy it. Paste it on a new Maple input line.

   > (1 - exp(x)) / (1 + exp(-x)) + (1 + exp(x)) / (1 - exp(-x));

4. Edit the line by wrapping the simplify command around the line you have copied. Use the *simplify* command.

   > simplify((1 - exp(x)) / (1 + exp(-x)) + (1 + exp(x)) / (1- exp(-x)));

   $$-4\frac{1}{(1+e^{(-x)})(-1+e^{(-x)})}$$

5. Copy this input line. You see from the output that Maple has not expanded the denominator. Maple will expand factors in the numerator, so you can force the expansion of the denominator by taking the reciprocal, expanding it, and taking the reciprocal again. Edit the line to make these changes and execute the command once more.

> 1/ (expand(1/ (simplify((1 - exp(x)) / (1 + exp(-x)) + (1+ exp(x)) / (1 - exp(-x))))));

$$\frac{1}{\frac{1}{4} - \frac{1}{4}\,\frac{1}{\left(e^{x}\right)^{2}}}$$

6. You want to evaluate the expression at a specific value of x. You can use the *subs* command on *any* of the previous lines since they are all equivalent. We will copy the first input line and substitute $x = 0.5$.

   > subs(x = 0.5, (1 - exp(x)) / (1 + exp(-x)) + (1 + exp(x)) / (1 - exp(-x)));

$$\frac{1 - e^{.5}}{1 + e^{(-.5)}} + \frac{1 + e^{.5}}{1 - e^{(-.5)}}$$

7. One final copy and paste using the *evalf* command finishes the problem.

   > evalf(subs(x = 0.5, (1 - exp(x)) / (1 + exp(-x)) + (1 +exp(x)) / (1 - exp(-x))));

   6.327906828

8. We can check our work by evaluating the simplified expression at $x = 0.5$. Just wrap *evalf(subs(x* = 0.5 . . .) around the command in step 5. The result is the same as in step 7.

In this whole process, you have input the mathematical expression only once, and you have checked Maple's formatted output to make sure that it matches the problem. After seeing that the expression you are working with is correct, you "wrap" commands around it and observe the results in a step-by-step process.

Do use Maple's Help facility. Type *?*, followed by the name of the topic. For instance, to get help on the substitute command, type *?subs* after the Maple prompt.

Do be careful about the decimal point (the period). Maple uses the same character for the decimal point and the concatenation[3] operator. Thus, you must be very careful with the syntax, as shown in the following examples:

Range specifier: You want the range to be 0 to .3, so you type:
0...3

Maple interprets this as 0 to 3! Since there is no space between the double dots and the single dot, it assumes that you meant two dots and a 3.

---

[3] See the help page. Type *?concatenation* or *?dot.*

The variable x has a value of 3 and you want to use 3.1, so you write

x.1,

thinking that Maple will substitute 3 for x and append the .1. Instead, Maple thinks you mean the variable *x1*.

Do be careful about the range specifier (two periods, one after the other). There must be no space between the periods in this case.

Don't write *sin(x)* as *sin x*, *log(x)* as *log x*, or *exp(x)* as *e^x*. (Even worse, don't write *sin*x* for *sin(x)*). Maple uses a consistent "functional" notation. A function depends on its variables. For instance, the *sin* function depends on the angle at which the sine is required. If the angle is theta, you write *sin(theta)* to evaluate the function. The expression *e^x* looks legal, and it is. The only drawback is that Maple treats *e* as the letter *e*, not as the base of the natural logarithms; thus, in Maple, *e* has no predefined value.

Do practice wrapping functions around other functions. This is called "nesting." A common use of this technique is finding the decimal approximation for some quantity. Say you want to find the decimal value for the square root of 2. You could issue two Maple commands:

> A := sqrt(2); evalf(A);

or you could accomplish the same thing in one command by typing the command:

> evalf(sqrt(2));

An advantage of this technique is that you will acquire the habit of "thinking one step ahead."

Do execute all the Maple input lines when you open a Maple worksheet. When you use the File, Open command to use a previously stored worksheet, Maple does not recalculate the commands in the worksheet unless you tell it to. The file may contain Maple output, but that was the output that was stored with the file. It is not freshly computed. Thus, any names that appear on Maple input lines will not be known in the current Maple session. You must reexecute these commands to have these names defined. There are two ways of reexecuting Maple commands once you open a worksheet.

1. In Release 4, use the mouse and select Edit, Execute, Selection after highlighting the command you want executed. In Release 3, use the Format, Execute Worksheet command. This causes Maple to execute every command in the worksheet. This is the easiest way to make sure that every name used in the worksheet is defined. Also, the output lines will be recomputed and will replace whatever was in the original file. The drawback with this method is that it may take a long time to recompute every command if the worksheet is large or contains commands that take a lot of computing.
2. Place the cursor on the first Maple input line and press *Enter (Return)*. Maple will execute the command and place the cursor on the next Maple command line. There

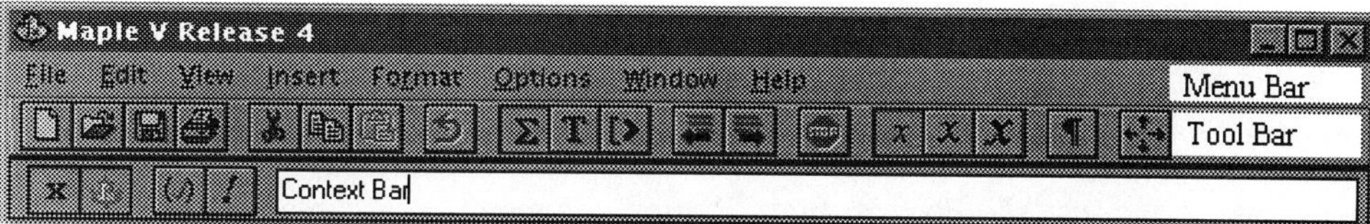

Figure A.4 The Menu Bar, Tool Bar, and Context Bar in Maple V, Release 4

may be text or possibly plot outputs or other graphics between the two command lines. Maple will skip right over the text and graphics and place the cursor on the next command line. Don't be surprised when this happens. There may be more than one screen of text, so the first command and its output will not be visible. If you use the vertical scroll bar to go back to the first command, you will see it and its output. It may not look any different, but it has been recomputed. Now for another surprise: if you press *Return*, the screen will jump once more! Because you used the vertical scroll bar, the cursor remained where it was, on the second Maple command line. Pressing *Return* caused Maple to execute the second command—write the output of the command and go on—placing the cursor at the third command line in the file. A further complication is: Maple allows "program groups," so that what was described as a single command may actually be a group of commands, depending on how the file was produced in the first place. The point to remember is that you can move through the worksheet, once you place the cursor on the first command, simply by pressing *Return* again and again.

3. You can be selective about which commands to run. Place the cursor on any Maple command line you wish to execute. Press *Return* and Maple will recompute the command line, or group of lines. It then places the cursor on the next command. You may execute that command if you wish, or you can move the cursor to some other command. This way, you can define only those names that you need for your current purpose. This last method may save you some time, but remember that you *must* execute the command lines if you want to use the names that have been mentioned in the worksheet. Just because you can see the names on the screen doesn't mean Maple knows what they are!

Using the Text Editor and Saving Your Worksheets

Maple has a built-in text editor (see Figure A.4) which you can use in describing your calculations and in the preparation of reports. In Release 4, you may select the text mode by clicking on the [T] button on the Tool bar (see Figure A.4). When you are finished typing text, you switch to a new input region by clicking on the [>] button.

In Release 3, you use the Format, Input Region command. You can also use a keyboard shortcut. Type F5 to switch between text input and command input.

To sum up, you can prepare a document containing text, Maple input lines, and Maple output and graphs, and you can save your worksheet on a disk. Saving to disk is accomplished either using the File, Save command, or by clicking the button in either Release 4 or Release 3.

Test Yourself

Answer these questions about Maple.

1. What are the syntax errors in these input lines? The parenthetical comment lines give what the Maple user intended.

(a) > eq1 = a*x^2 + b*x * c = y; Answer: ____________

(b) > eq2 := a*x^2 + b*x +c := y; Answer: ____________

(c) > eq3 := ax^2 +bx + c = y; (Quadratic formula) Answer: ____________

(d) > eq4 := y = 1/1+x; (Reciprocal of 1 plus *x*) Answer: ____________

(e) > eq5 := R*a, eq6 := P = V^2/R; Answer: ____________

(f) > eq7 := SQRT(16); (Square root function) Answer: ____________

(g) > eq8 := plot(x^2, x=0...5); (Plot x^2 from 0 to 0.5) Answer: ____________

2. Answer these questions assuming that each group begins a new Maple session. In other words, assume that the *Restart* command has been given at the start of each section. Predict the results of each set of commands. Read the input line very carefully. These questions have been designed to illustrate subtle differences between pairs of similar-looking constructs in Maple.

(a) > a = 3; a + 3; Answer: ____________

(b) y = m*x + b; subs(m = 3, b = –2, y); Answer: ____________

(c) y := m*x + b; subs(m = 3, b = –2, y); Answer: ____________

(d) A := Pi * r^2; subs(r = 2, A); Answer:________________

(e) A := pi * R^2; subs(R = 2, A); Answer:________________

(f) y := 1/3 + 1/4; z := 1./3 + 1/4; Answer:________________

3. State which commands are used to compute each of the following:

Example: Find all the factors of 1,155. Answer: ifactor (1,155)

(a) The exact square root of 65. Answer:________________

(b) The decimal approximation to the square root of 65. Answer:________________

(c) The hypotenuse, base = 7, height = 14. Answer:________________

(d) Solve the equation $3 + 4x = 5 + 6x$. Answer:________________

(e) Plot the line $y = 3x + 4$, for x from –2 to +1. Answer:________________

(f) All the factors of $6x^2 + 5x - 6$. Answer:________________

(g) Reduce $(x^2 - 1)/(x + 1)$ to simplest form. Answer:________________

(h) Plot the line segment from (2, 3) to (3, 5). Answer:________________

(i) Plot the curve $y = x^2$ for x from –2 to 2. Answer:________________

CHAPTER 1

Measuring Angles and Arc Lengths

Objectives for This Chapter

1. Learn the definitions for the parts of an angle
2. Use radian measure as an alternative to measuring angles in degrees
3. Apply the formula for arc length to the measurement of circular arcs
4. Use Maple to convert from radians to degrees and from degrees to radians

Maple Commands Used in This Chapter

deg_	A name used in this chapter for a simple way to convert to radians.
evalf(number)	Express a number as an approximate decimal.
rad_	A name used in this chapter for a simple way to convert to degrees.
subs(r = 3, theta = 4, s=r*theta)	Substitute $r = 3$, theta = 4 in the equation $s = r$ * theta.

Plane Angles

An angle is formed by two intersecting half lines, as shown in Figure 1.1.

Imagine the line called the *initial side* being rotated about point O (called the vertex), until it takes up the position of the *terminal side*. This rotation sweeps out an *angle*. Look at Figure 1.1(a). As line OA rotates around point O to take up position OB, an angle is formed. If you

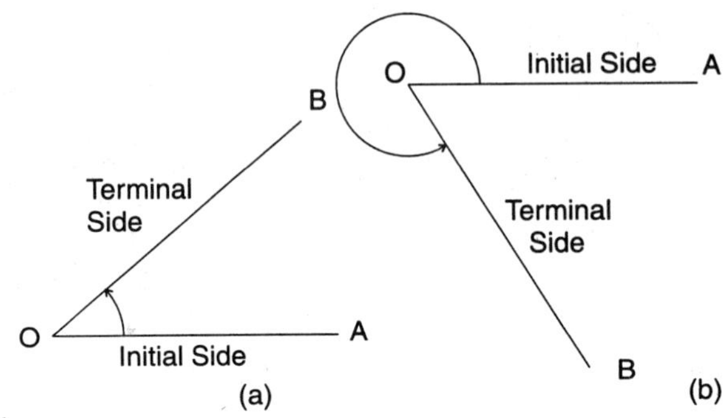

Figure 1.1 Angles

further imagine the line continuing to rotate until it takes up the position OB as shown in Figure 1.1(b), it has swept out a bigger angle. If the line kept rotating until it lay over its original position, it would have traced out a full revolution. We need a way of measuring angles other than as full rotations. There are two common units in use today. One unit of measure, the *degree*, divides the full revolution into 360 equal parts, and the second, the *radian*, divides the full revolution into 2π parts.

The Right Angle

For many centuries and indeed until very recently, angles were discussed only in relation to geometry. Geometers of old measured their angles in relation to the angle that is formed when the circle is divided into four equal, pie-shaped sections.

Such an angle is called a *right angle* (see Figure 1.2). A *degree* is an angular measure that divides the right angle into 90 equal parts; thus, a full circle subtends an angle of 360° at its center. In the older way of measuring angles, a *minute* is 1/60 of a degree, and a *second* is 1/60 of a minute. This system of measuring angles is rarely used today. Instead, the decimal notation is used. For example:

a) $36^\circ\ 24'30'' = 36 + 24/60 + 30/3600 = 36.4083333^\circ$

b) $1/2$ of $21^\circ\ 15' = 1/2$ of $21.25^\circ = 10.625^\circ$

c) $5 \text{ x } 7^\circ\ 49' = 35^\circ\ 245'' = 39^\circ\ 5'' = 39.08333^\circ$

d) $5 \text{ x } 7^\circ\ 49' = 5 \text{ x } 7.81666^o = 39.08333^\circ$

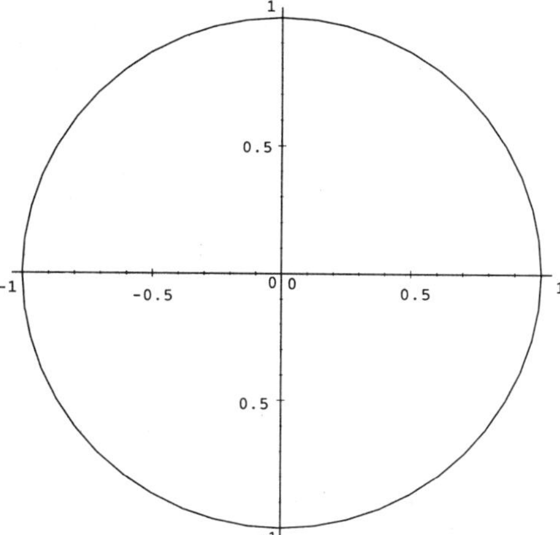

Figure 1.2 Four Pie-Shaped Sections of the Circle: Four Right Angles Are Formed

Radian Measure and Arc Length

In science and technology, the common unit of angular measure is the *radian*. By definition, the angle of a full rotation is 2π radians. This may seem like a poor choice for a unit, because a full rotation is not an even number ($2\pi = 6.283185\ldots$, so there are 6.283185 . . . radians in a full circle). On the other hand, many formulas are made simpler by measuring angles in radians. Consider the problem of finding the length of a circular arc. You know that the circumference of a circle of radius 1 is 2π. Also, there are 2π radians in a full revolution. A little reflection about linear variation or proportional behavior will convince you that you can write the formula for the length of any circular arc as

$$s = r\theta,\ \theta \text{ in radians} \qquad (1\text{-}1)$$

You can prove to yourself that the formula must be correct by observing that, for any size circle, the arc length s must be proportional to the radius, because if you double the radius, you double the length of the circular arc as well. The same is true for the angle at the center of the circular arc. If you double this angle, you double the length of the arc. Since the formula works for a radius of 1 and a full revolution, it must work for any angle and any radius because of these proportionalities (See Figure 1.3).

Example 1-1

What is the length of a circular arc if the radius is 30 centimeters (cm) and the angle at the center is 1 radian?

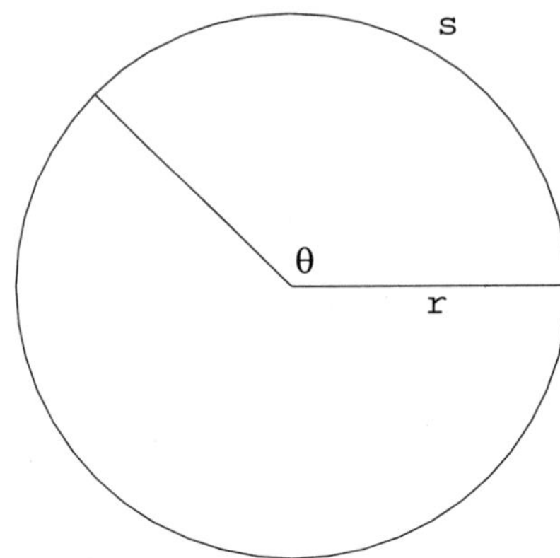

Figure 1.3 The Formula for Arc Length

Solution: The length is $s = r\,\theta$, $s = 30 \times 1 = 30$ cm.

Example 1-2

What is the length of a circular arc if the radius is 1.56 meters (m) and the angle at the center is 66°?

Solution: First convert the angle to radians (rad): $66° = 66 \times \pi / 180 = 1.152$ rad. Thus, $s = r\,\theta$, $s = 1.56 \times 1.152 = 1.797$ m = 1.8 m.

Example 1-3

You can use Maple to compute arc length even if the central angle is given in degrees. Since this example involves converting degrees to radians, see example PP1–2.

Your Turn: (a) What is the length of a circular arc if the radius is 472 m and the angle at the center is 1.818 radians?

Answer: ______________________________

(b) What is the length of a circular arc if the radius is 25.4 cm and the angle at the center is 23°?

Answer: ______________________________

(c) What is the length of a circular arc if the radius is 254 cm and the angle at the center is 90°? Do you need the formula for arc length to solve this?

Answer: ______________________________

Converting between Radians and Degrees in Maple

If you expect to have to convert back and forth between radians and degrees for a set of problems you are working on, it is useful to assign some conversion factors in a Maple session. This is accomplished by defining the names rad_ and deg_ . Both of these names have the underscore character (SHIFT+_) as their last letter. You should never use the underscore character as the first letter in any name you define, because Maple itself uses names beginning with an underscore, but there is no restriction about using the underscore character as the last letter in any name you define. Using the underscore may help you to remember that this is a quantity you defined in order to help you convert units. Here are the definitions:

> rad_ := 180/Pi*deg;

$$\text{rad\_} := 180\frac{deg}{\pi}$$

> deg_ := Pi/180*rad;

$$\text{deg\_} := \frac{1}{180}\pi rad$$

Examples:

a) Convert 3/5 radians to degrees. *Solution:*

> 3/5*rad_;

$$108\frac{deg}{\pi}$$

b) Convert 0.45 radians to degrees and express as a decimal. *Solution:*

> evalf(0.45*rad_);

```
25.78310077 deg
```

Here, the *evalf* function is used to convert the answer to decimal form.

c) Convert 45° to radians. *Solution:*

> 45*deg_;

$$\frac{1}{4}\pi rad$$

d) Convert 33° to radians, in decimal form. Solution:

> evalf(33*deg_);

```
.5759586531 rad
```

e) Use Maple to define a formula for arc length and use it to find the length of an arc when the radius is 5 cm and the angle at the center is 135°. Solution: First define the formula.

> s := r * theta/rad;

$$s := \frac{r\theta}{rad}$$

Then use the substitute command to supply values for *r* and θ.

> subs(r=5*cm, theta= 135*deg_, s);

$$\frac{15}{4}cm\pi$$

One drawback is that Maple writes π last instead of writing the units (cm) last, in the preferred way. This defect is not present if we convert to decimal notation:

> evalf(subs(r=5*cm, theta= 135*deg_, s));

```
11.78097245 cm
```

Notice that the answer is given as a numerical quantity and its associated unit. Since the names *deg* and *rad* are used as part of the definition of the conversion factors, you must not use them for anything else in the same worksheet. If you might have used these as variable names earlier in the Maple session, you should issue these two commands before attempting to define *deg_* and *rad_*:

> deg := 'deg'; rad := 'rad';

This will clear any values these names may have had previously. Of course, the same holds true for the name *cm*, which is used in the formula for arc length.

Paper and Pencil Exercises

PP1-1

(a) Label the diagram in Figure 1.4. Show the location of the initial side, terminal side, and vertex.

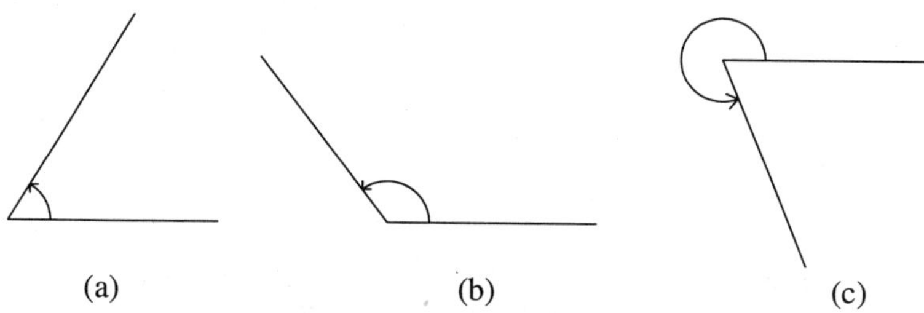

Figure 1.4. Diagram for Problem PP1-1

(b) Use the protractor to measure the angles.

PP1–1a: ____________

PP1–1b: ____________

PP1–1c: ____________

(c) Convert these angles to radian measure. Mentally compare the magnitudes of the angles in radians to the same angles measured in degrees.

PP1–1a: ____________

PP1–1b: ____________

PP1–1c: ____________

PP1–2

Convert these angles to radians. Answers are given for every fourth problem so that you can check your work.

(a) 287 degrees — *Answer:* 5.01 radians

(b) 355 degrees — *Answer:* ____________

(c) –3 degrees — *Answer:* ____________

(d) –160 degrees — *Answer:* ____________

(e) –97 degrees — *Answer:* –1.69 radians

(f) –269 degrees — *Answer:* ____________

(g) 284 degrees — *Answer:* ____________

(h) –303 degrees — *Answer:* ____________

(i) 80 degrees *Answer:* 1.40 radians

(j) –57 degrees *Answer:* ____________

PP1–3

Convert these angles to degrees. Small angles are given in Maple's exponential notation (*e*– = 1/10 = 0.1, *e*–2 = 1/100 = 0.01, etc.)

(a) –5.17 radians *Answer:* –296 degrees

(b) –1.75 radians *Answer:* ____________

(c) .800e–1 radians *Answer:* ____________

(d) –3.04 radians *Answer:* ____________

(e) –3.54 radians *Answer:* –203 degrees

(f) –2.65 radians *Answer:* ____________

(g) –2.32 radians *Answer:* ____________

(h) 2.87 radians *Answer:* ____________

(i) 4.05 radians *Answer:* 232 degrees

(j) 2.41 radians *Answer:* ____________

(k) –4.21 radians *Answer:* ____________

(l) –5.27 radians *Answer:* ____________

(m) –.340 radians *Answer:* –19.5 degrees

(n) –.670 radians *Answer:* ____________

(o) 2.74 radians *Answer:* ____________

(p) 1.84 radians *Answer:* ____________

(q) –.830 radians *Answer:* –47.5 degrees

(r) 5.73 radians *Answer:* ____________

(s) –.650 radians *Answer:* ____________

(t) .290 radians *Answer:* ____________

PP1–4

Convert these angles to degrees. The angles are given in terms of π, since radian measure is frequently reported this way. Convert the radian measure to a pure decimal number by multiplying by 3.18 before converting to degrees.

(a) 1.69 π radians *Answer:* 304 degrees

(b) 2.51 π radians *Answer:* ____________

(c) 1.25 π radians *Answer:* ____________

(d) .890 π radians *Answer:* ____________

(e) –.440 π radians *Answer:* –79.2 degrees

(f) 2.60 π radians *Answer:* ____________

(g) 1.01 π radians *Answer:* ____________

(h) 1.59 π radians *Answer:* ____________

(i) –.610 π radians *Answer:* –110 degrees

(j) .720 π radians *Answer:* ____________

(k) 2.39 π radians *Answer:* ____________

(l) 2.72 π radians *Answer:* ____________

(m) –.850 π radians *Answer:* –153 degrees

(n) 2.20 π radians *Answer:* ____________

(o) 1.63 π radians *Answer:* ____________

(p) –.710 π radians *Answer:* ____________

(q) 2.94 π radians *Answer:* 529 degrees

(r) –.780 π radians *Answer:* ____________

(s) –.510 π radians *Answer:* ____________

(t) 2.91 π radians *Answer:* ____________

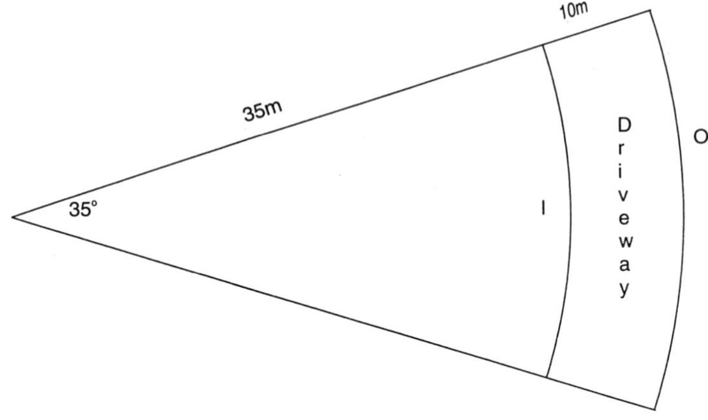

Figure 1.5 Diagram for Problem PP1–5

PP1–5

A driveway is in the form of a circular arc (see Figure 1.5). What are the inner and outer lengths of the arcs? The arcs are labeled *I* and *O* on the diagram.

Convert the angle to radian measure: *Answer:* ________________

Length of the inner arc, *I*: *Answer:* ________________

Length of the outer arc, *O*: *Answer:* ________________

PP1–5

Roughly, to two significant figures, how many degrees are there in one radian?

Answer: ________________________________

PP1–6

What is the formula for arc length? Identify the terms and state the units.

Answer: ________________________________

PP1-7

Let us say that you have defined the calculation for arc length as the Maple command:

> s := evalf(r * 0.174532925);

$$s := .174532925r$$

Write the Maple command that will find the distance s by substituting the values 311 m for r and 75° for θ. Calculate s without using Maple.

> ; (Write the Maple command here) *Answer:* ____________________

PP1-8

How do you substitute for s in the above command? Careful! Is this possible to do?

Answer: ____

PP1-9

If you write in Maple:

> s := r * theta;

$$s := r\theta$$

the variable s is a *name* whose *value* is $r\theta$. What happens if you write:

> s = r * theta; *Answer:* ____________________

Maple Lab

ML1-1

Express these radian measure angles in degrees. Work out the answers using your calculator and verify the results using Maple. Use the definitions of this section in your work.

(a) 0.123 r = __________ degrees (b) 0.0174532 r = __________ degrees

(c) 0.5236 r = __________ degrees (d) 1.0472 r = __________ degrees

(e) 6.2832 r = __________ degrees (f) 100 r = __________ degrees

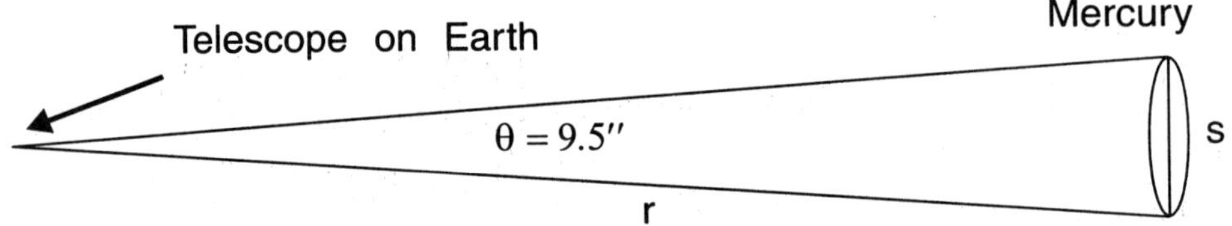

Figure 1.6 Angle of Mercury, Measured by an Observer on Earth

ML1–2

The planet Mercury subtends an angle of 9.5″ as measured by a telescope on earth when the distance from earth to Mercury is 65 million miles. What is the diameter of Mercury from these measurements? (See Figure 1.6.)

Answer: The diameter of Mercury is ________ miles (don't forget to convert the angle to radians).

ML1–3

Scientists at the National Aeronautics and Space Administration (NASA) shine a powerful laser beam on the moon from a site on Earth. When it reaches the moon, the diameter of the beam is 25 kilometers (km). The divergence of the laser beam is 6.5×10^{-5} radians. Use this information to calculate the distance from the earth to the moon. Hint: the diagram for this problem is the same as for problem ML1–2, but this time, *r* is the unknown. The Maple way of writing 6.5×10^{-5} is 6.5*e*–5.

Answer: The distance from the earth to the moon is ________ km.

ML1–4

Start a fresh Maple session. Type in the commands:

```
> rad_ := 180/Pi*deg;
> deg_ := Pi/180*rad;
```

You are ready to convert from degrees to radians and vice versa. Use these Maple commands to convert:

(a) 0.125 radians to degrees. *Answer:* ________________

(b) π degrees to radians. *Answer:* ________________

ML1–5

If you use the definitions in ML1–4, Maple will leave π as a symbolic value rather than evaluating it as a decimal. Redefine the conversion process by typing in these commands:

> **rad_ := 57.29577951*deg;**

> **deg_ := 0.0174532925*rad;**

Use these commands to convert:

(a) 6.2831853 radians to degrees *Answer:* ____________

(b) 114.59159 degrees to radians *Answer:* ____________

ML1–6

Type these commands and observe the results (i.e., the Maple output):

> **a := evalf(d +m/60 + s/3600);**

Use this command and the *subs* command to convert from degrees, minutes, and seconds to decimal degrees.

> ; (Write the Maple substitute command here)

(a) Convert 44° 35′ 24″ to decimal degrees. *Answer:* ____________

(b) Convert 89° 59′ 59″ to decimal degrees. *Answer:* ____________

ML1–7

Now that you have both commands from ML1–6 available, do some more conversions. The command that gives the name *a* a value does not need to be changed. Edit the *subs* command to convert 43° 36′ 25″ to decimal form. Simply highlight the numbers you want to change and put in the new values.

Answer: ____________

ML1–8

There is another method, which allows you to keep previous results on the screen. Highlight the whole command and copy it. Paste it on a new command line, and then edit the values you want to change. Use this technique to convert the angles in (a) through (c):

(a) 10° 10′ 10″ *Answer:* ____________

(b) 20° 30′ 40″ *Answer:* ____________

(c) 53° 35′ 53″ *Answer:* ____________

ML1-9

One of the most useful things you can learn by using Maple is how to wrap one command around another. The process can be continued by wrapping yet another command around the first two, and so on. By practicing this technique, you are developing your ability to do multistep problems. In this "nested function" approach, the outermost layer is the answer you want for the problem at hand, while the innermost layer is the first thing you have to do to find that answer. The present problem illustrates the technique. How would you accomplish the conversion of an angle, given in decimal degrees, to degrees, minutes, and seconds? Here is a partial solution, using 24.51° as an example.

1. Assign the value to a variable (a name):
2. > **a := 24.51;**
3. The integral part will remain the same, so you need to begin with the fractional part. You could define *m* as
4. > **m := 60*(a–24);** (Compute the fractional part of 24.51, which is 0.51)
5. Find the number of seconds by evaluating the fractional part found in step 4.
6. > **s := 60 * (m-30);**

You can put all this together in one command. The step-by-step process has been laid out. There are a couple of Maple commands that you will need. The *frac* command extracts the fractional part of a number, and the *trunc* command picks out the integer part. Examine the following command carefully:

> **a := 'a';** (Make sure *a* does not have an assigned value)

> **a1 := trunc(a-frac(a)), trunc(60*frac(a-trunc(a))), trunc(60*frac(60*(a-trunc(a))));**

```
a1 :=trunc(a - frac(a)), trunc(60 frac(a - trunc(a))),
          trunc (60 frac(60a - 60 trunc(a)))
```

Now issue the command:

> **a := 24.51; a1;** (Convert *a* to degrees, minutes, and seconds)

Use this command to convert (a) 17.99°, (b) 1.0173°, (c) 12.24° to degrees, minutes, and seconds.

Answers: (a) ________ (b) ________ (c) ________

ML1–10

Explain how you would convert the sum of 30° 45′ 55″ and 59.23472° to degrees, minutes, and seconds.

Answer: ______________________________

ML1–11

Write the Maple formula that converts the angle 75° 13′ 11″ (given in degrees, minutes, and seconds) to radians.

Partial Solution. Let *d* be the integer part, *m* the minutes, and *s* the seconds. Start with the assignment:

> d := 75; m:= 13; s := 11;

> ; (Write the Maple command here) *Answer:* ______________

ML1–12

You could have converted ML1–11 in two ways; using the exact conversion or the decimal approximation. Give the *other* solution here:

> ; (Write the Maple command here) *Answer:* ______________

ML1–13

A beam of light enters a region between two mirrors at a small angle. As it is reflected back and forth, the beam spreads. How will it spread after 15 reflections, if a) the mirrors are 1 meter (m) apart and (b) the divergence of the beam is 0.1 radian?

Method of Solution. You can solve this problem by the same method as the previous two, but you need to draw a diagram. Of course, you can let Maple draw the diagram for you! Enter the following commands in a Maple session. These commands contain a *Maple program* (also called a Maple procedure). The procedure (which you need not analyze in detail), creates a plot for this problem. You can measure the divergence of the beam directly on the plot using the mouse. The size of the beam is indicated between the red and green lines on the graph. The beam of light enters the mirror region at the bottom left, at coordinates (0, 0). As the beam progresses, it spreads and it is reflected between the mirrors. After 15 reflections, the beam will be at the left-hand mirror (the *y* axis), at a height of 8 units. The red line represents the lower edge of the beam, and the green line represents the upper edge. Measure the divergence of the beam on the Maple plot. Calculate the divergence of the beam and compare this to the measurement on the graph.

Enter these statements into Maple:

```
> bounce := proc(n, d, d1)
> local i, l, m, p1, p2, p3;
> l := 0, 0; m := 0, 0;
> for i from 1 to n do
> l := l, irem(i, 2), i*d/2;
> m := m, irem(i,2), i*(d+d1)/2;
> od:
> l := [l]; m:= [m];
> p1 := plot( { [[0, -0.1],[ 0, 10], [1, -0.1], [1, 10], [0, 10], [1, 10]] }, -0.1 .. 1.1, color =
black):
> p2 := plot( l , -0.1 .. 1.1, color = red):
> p3:=plot( m , -0.1 .. 1.1, color = green):
> plots[display]( { p1, p2, p3 }, title = `Reflections` );
> end:
> bounce( 16, 1, .1);
```

Answer: The divergence of the beam after 15 reflections is _________ m.

CHAPTER 2

Review of Algebra

Objectives for This Chapter

1. Review the difference between a whole number, a fraction, and a real number
2. Understand the difference between an exact number and its decimal approximation
3. Know how to convert exact numbers to approximate decimals using Maple
4. Review how polynomials in one or two variables are expressed in factored or expanded form
5. Know how to use Maple to factor or expand polynomials
6. Use Maple to substitute values for variables in expressions
7. Use Maple to plot the graph of a function or expression
8. Solve linear equations using Maple

Maple Commands Used in This Chapter

*	The multiplication operator. Example: 2 * 3
/	The division operator. Example: 3 / 2
^	The power operator. Example: 2^3
Pi	The number Pi. The area of a circle of radius 1.
evalf(expr)	Convert to a decimal value. Example: evalf(Pi)
expand(expr)	Expand an expression by multiplying out the terms.

factor(expr)	Factor an algebraic polynomial.
igcd(i1, i2)	The greatest common divisor of two integers. Example: igcd(16, 24)
isolve(eqn)	Find integer solutions for the variables in an equation.
plot(expr, range)	Plot an expression over a specified range. Example: plot(sin(x), x = 0 .. Pi)
solve(eqn, var)	Solve an equation for the specified variable. Example: solve(x = 3*x–8, x)
sqrt(number)	The square root of a number. Example: sqrt(25)
subs(eqn, expr)	Substitute a value for one of the variables in an expression. Example: subs(x =2, x^2)

Problem Solving, Step by Step

The practice of trigonometry requires that you be comfortable with algebraic operations. Those of most use will be presented in this chapter. Even if you know algebra, it will be useful for you to scan the material covered here because you will familiarize yourself with the Maple expressions that correspond to common algebraic operations. Here are the steps to follow:

1. Understand the concept. In other words, know how to do the problem by "paper and pencil."
2. Read the Maple input line to see how a problem is expressed in Maple notation.
3. Study the Maple output. Observe what Maple has done with the input line and decide whether it solves the problem or, at least, furthers a computation.
4. Put it all together. Review the problem statement and the steps leading to the solution. Ask yourself, "If I were presented with a similar problem, what is the step-by-step procedure I would employ to solve it?" Then mentally go through the steps. This review process is essential to solidify the understanding you have gained into your long-term memory.

Fractions

The idea of a fraction arises when you think of dividing a whole into some number of parts. For example, you could imagine slicing an apple pie into eight equal parts. Then each slice would be 1/8th of the whole pie. Three slices are 3/8 of a pie, but two slices are 1/4 of a pie.

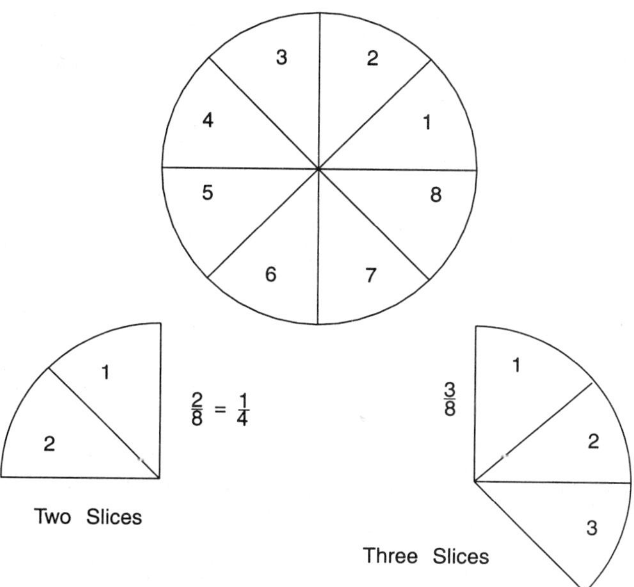

Figure 2.1 Fractions Are Created When a Whole Is Divided into Parts

You can see in Figure 2.1 what 3/8 and 2/8 of a pie look like. Also, just by looking at the shape, you see that two slices make 1/4 of a pie. Four of these shapes make a full pie. Numerically, we have:

$$\frac{2}{8} = \frac{2 \times 1}{2 \times 4} = \left(\frac{2}{2}\right) \cdot \left(\frac{1}{4}\right) = \frac{1}{4} \tag{2-1}$$

No doubt, you are quite familiar with the technique of taking a fraction, like 2/8, and finding its simplest form, 1/4. The problem of finding the simplest form is called reducing the fraction to lowest terms.

From examples like this one, you can generalize the concept of a fraction. The definition of a fraction is: the ratio of two integers, where the denominator must not be zero. In symbols:

$$\mathrm{a} = \frac{p}{q}, q \neq 0 \tag{2-2}$$

Thus, 3/4, 9/10, 355/17, and 24/24 are all fractions, even though the last, 24/24, is not expressed in the lowest terms. The following are not fractions: $\sqrt{2}/3$, $22/\pi$. Both $\sqrt{2}$ and π

are *irrational* numbers. These numbers are not fractions, so any term containing π or $\sqrt{2}$ cannot be a fraction.

Since Maple tries to work as accurately as possible, it often gives answers as fractions. For instance, if you type:

```
> 55/5;
```

```
11
```

Maple responds with the number 11. The number 5 divides the number 55 evenly, 11 times. On the other hand, if you type

```
> 22/7;
```

$$\frac{22}{7}$$

Maple simply repeats the question! Why doesn't Maple respond with 3.1428? The answer lies in the way some fractions are represented by decimals. Take the number 1/3 and express it as a decimal:

$$1/3 = 0.3333333333333333333333333333333333333333333333333333333333\ldots$$

You see that the decimal representation of 1/3 is a list of 3s which goes on forever. Since Maple tries to be as accurate as possible, it leaves its answer in fractional form. Of course, it is often desirable to express answers in decimal form, and Maple has a command to accomplish this. The command is called *evalf*, and it works like this:

```
> evalf(22/7);
```

```
3.142857143
```

Whenever you want Maple to convert something to decimal form, you must "wrap an *evalf* around it." You type in *evalf*, and you *enclose the expression to be evaluated in ordinary parentheses.* No other brackets will do, as Maple uses the brackets [ ] and { } for other purposes. There is one other form of the *evalf* command, however. It is used if you want to evaluate your answer to more than the 10 figures that Maple routinely gives. (It's not a good idea to use less than 10 figures. Instead, you should take the 10 figures Maple gives you and round them down yourself.) Here are some examples.

Example 2-1: A Large Fraction

There are 1,304,903 square miles (sq. mi.) in Canada's Northwest Territories. The area of Texas is 267,339 sq. mi. Comparing the two areas, what fraction of the Northwest Territories is Texas? Solution: use the area of Texas as the numerator and the area of the Northwest Territories as the denominator to form a fraction.

> **267339/1304903;**

$$\frac{267339}{1304903}$$

Since Maple gives the same answer as the input, you know that the two numbers, 267,339 and 1,304,903, have no common factors. We can always express this fraction as a percentage, namely:

> **100*evalf(267339/1304903)*`%`;**

```
20.48726994%
```

We had to resort to some fancy trickery to get the percent sign in, but basically, this is the *evalf* function applied to the fraction. (The whole expression had to be multiplied by 100 to convert it to percent.)

Example 2-2: The Number of People per Square Mile in New York State

The population of the state of New York in 1986 was 18,194,025, and the area of the state is 49,575 sq. mi. How many people per square mile does this represent? Solution:

> **18194025/49575;**

```
367
```

Maple reduces this fraction to a whole number because 49,575 divides 18,194,025 exactly.

Two related commands are *ifactor* and *igcd.* The *ifactor* command will give you the factors of an integer. Factoring the numbers from the previous problem gives

> **ifactor(18194025); ifactor(49575);**

$$(3)\,(5)^2\,(367)\,(661)$$

$$(3)\,(5)^2\,(661)$$

The common factors are 3, 5, 5 and 661, as shown by the *igcd* function. The command *igcd* gives the greatest common divisor of two (or more) numbers:

> **igcd(18194025, 49575);**

```
49575
```

Your Turn

1. Find the factors of 281,522,223,382,549.

Answer: The factors are ____.

2. Find the factors of 3,011,753,745.

Answer: The factors are ____

3. What is the greatest common divisor of 2,345 and 5,670?

Answer: The greatest common divisor is _______________________________.

4. Add (or subtract) the following fractions (leave the answers as fractions):

   a) 344/45 + 67/90 + 37/90 *Answer:* ________________

   b) 1/2+2/3+3/4+4/5+5/6+6/7+7/8+8/9+9/10 *Answer:* ________________

   c) 1/4–1/8–2/15 *Answer:* ________________

   d) 2/9–5/11–2/17 *Answer:* ________________

5. Is the following correct? $\frac{2\not{6}}{\not{6}5} = \frac{2}{5}$ *Answer:* ________________

Real Numbers

Real numbers play an important role in trigonometry that fractions cannot fulfill. Just about the simplest triangle you can construct is attained by drawing a diagonal in a square whose side has a length of 1 (Figure 2.2).

The diagonal line is called the hypotenuse of the right-angled triangle. What is the length of this hypotenuse? In the diagram, this length is labeled *H*. Applying Pythagoras' theorem to this triangle, we find:

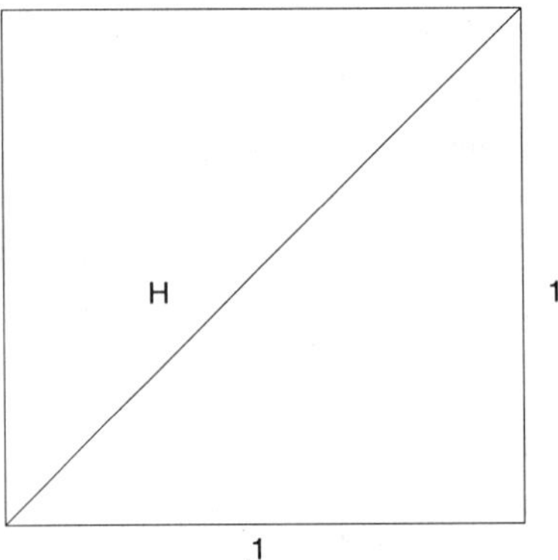

Figure 2.2 The Triangle Formed by Drawing a Diagonal in the Unit Square

$$1^2 + 1^2 = H^2$$

$$H^2 = 2, H = \sqrt{2} \tag{2-3}$$

The square root of 2, $\sqrt{2}$, is not a fraction, but it is a real number. Because numbers like $\sqrt{2}$, $\sqrt{3}$, $\sqrt{5}$, and so on can appear as trigonometric lengths, they must be part of our number system, yet they are not fractions. These numbers are called "irrational" to distinguish them from fractions, which are also called rational numbers. Combining the rational numbers and the irrationals, we arrive at the real number system.

Here are some ways you can operate with real numbers in Maple.

You can evaluate an irrational number, for example:

> sqrt(2), evalf(sqrt(2));

```
√2, 1.414213562
```

You can test whether a number is irrational or not:

> is(sqrt(2), real), is(sqrt(2), rational);

```
true, FAIL
```

You can evaluate irrational numbers to any "reasonable" number of decimal places:

> evalf(Pi, 60);

```
3.14159265358979323846264338327950288419716939937510582097494
```

Evaluating an answer using *evalf* will allow you to give answers in decimal form.

Example 2-3: Solving a Right Triangle

The hypotenuse of a right-angled triangle is 13.5 centimeters (cm) long. The base is 5.75 cm. What is the height?

Solution. Draw a sketch and label the diagram with the given information (Figure 2.3).

Now that you can visualize the problem, you know that the length you are looking for is more than 5.75 cm and less than 13.5 cm. You can solve the command using Maple, and you can check it using your calculator. This is especially important while you are learning Maple. You will know your calculator better at first, so use it to keep yourself on the right track. Here is the Maple solution. Calling the hypotenuse H, the base b, and the height h, we have, by the Pythagorean theorem:

$$h^2 + b^2 = H^2$$

$$\therefore \quad h^2 = H^2 - b^2 \tag{2-4}$$

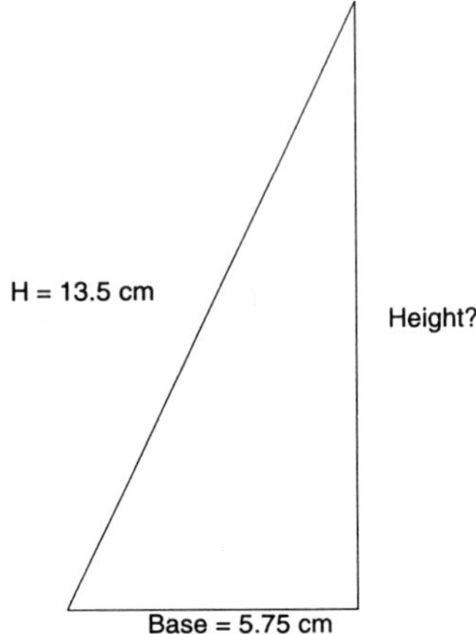

Figure 2.3 Solution of a Right-Angled Triangle

```
> h^2=13.5^2-5.75^2;
```

$$h^2 = 149.1875$$

```
> h = sqrt(13.5^2-5.75^2);
```

$$h = 12.21423350$$

You do not have to do the problem in two steps, first finding h^2, and then h. You need only type the command line shown. Even better, you can type in:

```
> h^2=13.5^2–5.75^2;
```

and evaluate it. Then you can wrap a *sqrt* around it by editing. Maple makes it easy to modify previously written commands for reexecution. The second-to-last step in any problem is to express the answer to a reasonable number of decimal places and give the units of the measurement. Since the hypotenuse was given to three-figure accuracy, we would state that the height of the triangle is 12.2 cm. The last (and most important) step is to verify the correctness of the answer. Maple makes this step easy. Go back to the formula for the base:

```
> h^2=13.5^2-5.75^2;
```

and edit it. Replace the term *h^2* with your answer:

```
> 12.2^2=13.5^2-5.75^2;
```

$$148.84 = 149.1875$$

You see that the answer has been verified to three significant figures. If you wanted more accuracy, you could replace *h* by using:

h = sqrt(13.5^2–5.75^2);

producing:

> sqrt(13.5^2-5.75^2)^2=13.5^2-5.75^2;

```
149.1875000 = 149.1875
```

Remember: verifying your calculations is simplicity itself when you use Maple, and it is also the most important step in any problem.

Your Turn:

1. Evaluate the square root of 2 to 30 places. Is the last digit even or odd?

Answer: ______________________________

Has the last digit been rounded? Would the 30th digit change if you evaluated to 33 places instead of 30 places?

Answer: ______________________________

2. Evaluate the fraction 1/7 to a large number of decimal places in order to find the repeating part. Convert this repeating part to a decimal and multiply it by 2, 3, 4, 5, 6, and 7. Let x equal the repeating part.

x = ________

2x = ________

3x = ________

4x = ________

5x = ________

6x = ________

7x = ________

Look closely at the digits for 2x, 3x, 4x, 5x, and 6x. What is unusual about them?

Answer: ______________________________

3. The Greeks were fond of a ratio that they called the Golden Section, and the Renaissance Italians called the *Divina Proporzione.* It is also called dividing the line into the "extreme and mean" proportion. The Golden Section is the number $\frac{\sqrt{5}+1}{2}$. Convert this number to a decimal:

Answer: ______________________________

4. In the next chapter, you will see that cos(x) is one of the basic trigonometric functions. For now, accept that there is a function called cos(x), which both Maple and your calculator recognize. Use Maple to calculate 2cos(π/5). Compare your result to the last answer.

Answer: ______________________________

5. Find the value of the continued radical $\sqrt{1+\sqrt{1+\sqrt{1+\sqrt{1+\ldots}}}}$

Using Maple, you type, *sqrt(1 +*, and you copy this part of the expression. Then you place the cursor at the end of the expression, and you type *CTRL+ V* to paste another one in. Type *CTRL+ V* a number of times, and then close the expression by typing, *1))))*. Use as many closing brackets as you have *sqrt* calls. For example:

```
> sqrt(1 + sqrt(1 + sqrt(1 + sqrt(1 + sqrt(1 + 1)))));
```

$$\sqrt{1+\sqrt{1+\sqrt{1+\sqrt{1+\sqrt{2}}}}}$$

Evaluate this expression using *evalf.* What is the answer?

Answer: ______________________________

Compare this answer to problem 3. Extend the number of *sqrts* in the continued radical and convert that to a decimal value. Does this approximate the Golden Section even more closely?

Answer: ______________________________

Algebraic Fractions and Factoring

The concept of a fraction can be extended to algebraic polynomials, an algebraic form that contains only positive exponents. Examples are:

$$\text{Monomial:}\quad 3x^3y^2 \tag{2-5a}$$

$$\text{Binomial:}\quad 4xy^2 + 12x^2y \tag{2-5b}$$

$$\text{Polynomial:}\quad x^3 + 6x^2y + 12xy^2 + 8y^3 \tag{2-5c}$$

You can use the substitute command, *subs,* to evaluate polynomials for given values of the variables.

Example: Evaluate the polynomial in Equation 2-5c if $x = 12$ and $y = 25$. *Solution:*

```
> subs(x=12, y=25, x^3 + 6*x^2*y + 12*x*y^2 + 8*y^3);
```

$$238328$$

Once you have typed the expression into Maple, you can edit it as many times as you wish. You could find the value of the expression for x=6 and y=3 simply by editing that part of the expression. Alternatively, you can copy the whole expression, paste it on a fresh Maple input line, and then edit it. This way, you keep a record of all the results in your Maple session.

Factoring Polynomials

Maple's routine for factoring polynomials is called *factor*. Here are a few examples:

```
> factor(x^3+6*x^2*y+12*x*y^2+8*y^3);
```

$$(x + 2y)^3$$

```
> factor( x^3-1);
```

$$(x - 1)(x^2 + x + 1)$$

```
> factor( 2*x^8-2);
```

$$2(x - 1)(x + 1)(x^2 + 1)(x^4 + 1)$$

You can see from these examples that Maple can help you examine whether any polynomial can be factored.

There is a related operation called *expand*, which multiplies out the factors. Examples of this command are:

```
> expand( (x-1)*(x+1));
```

$$x^2 - 1$$

```
> expand( (3*x+y)*(2*x-y)*(x-2*y)*(4*x+y) );
```

$$24x^4 - 46x^3y - 9x^2y^2 + 9xy^3 + 2y^4$$

```
> expand( sin(x+y) );
```

$$\sin(x)\cos(y) + \cos(x)\sin(y)$$

The last example takes us ahead of ourselves to show you that Maple can expand trigonometric as well as algebraic terms. There are many other facilities in Maple for manipulating polynomials, which are not covered in this brief review of algebra. Use the Help function in Maple or refer to the algebra book in this series (Parker: *Maple for Algebra*).

Your Turn

1. Factor (x^{90}–1). There are terms whose highest power of x is x^1, x^2, x^4, x^6, x^8, and x^{24}. How many of each are there?

Answer: There are __________ terms of each of these powers. The terms are all to the same degree in x, but they are not identical.

2. Factor $391x^2 - 440x - 1311$. *Answer:* ____________________

3. Expand $(1-x)(1+x)$; $(x+a)(x+a)$; $(bx+a)(bx+a)$.

Answers: ____

4. The following are called Special Products. Expand them and learn to recognize the simpler ones.

a) $(a-b)(a+b)$ *Answer:* ____________________

b) $(a+b)^2$ *Answer:* ____________________

c) $(x+1)^3$ *Answer:* ____________________

Functions and Their Graphs

The idea of a function is simple. A function (of one variable) takes a value you give it and it gives you back another value. That is, there is a one-to-one correspondence between the number you supply and the number you get back—the value of the function. The functions you will be examining will be expressible by a formula of the type:

$$y = f(x) \tag{2-6}$$

which in words is, "y is a function of x." The number you supply is x, and the number you get back is y. We say that x is the independent variable and y is the dependent variable. There is only one value of y for a given x in a true function, but there may be more than one value of x for a given y. An obvious example is the function $y = \sin(x)$ from trigonometry, which we will see in the next chapter. For our purposes in this section, the full power of Maple's functional notation will not be needed. Here, we want to show you how to evaluate functions and plot them.

You should know the following definitions:

- Domain: The set of values of the independent variable (x) for which the function is defined.
- Range: The corresponding set of values for the dependent variable (y).

- Maximum: The largest value of the function within the range, if it exists.
- Minimum: The smallest value of the function within the range, if it exists.
- Zeros: The set of values of the independent variable (x) for which $y = 0$.

A plot of a function gives you a picture. These pictures, called graphs, permit you to see the overall behavior of the function. Graphs are useful because the human mind understands pictures much better than tables of values or formulas.

Example 2-4: Graphs of Simple Functions

The Maple *plot* command requires two pieces of information, called *parameters* or *arguments*. These indicate the expression to be plotted and the range of the independent variable. You do not type in the equation, only the f(x) part. Thus, if you want to plot a function, you must express it in the form $y = f(x)$; then, you take only the f(x) part as the thing to be plotted. You must also give the *plot* command the range along the x axis over which the plot is to occur. Here is a sample of the *plot* command for the function $y = x^2$, where the part of the function to be plotted goes from $x = -2$ to $x = +3$ (see Figure 2.4). Note that a comma (,) separates the expression to be plotted and the range of the plot. Also, examine the range specification carefully. To plot from $x = -2$ to $x = +3$, you type x = -2 .. 3. A pair of periods, or "double dots," must be between the beginning point of the range and the terminal point. Of course, the beginning point of the range must be smaller than the end point. A final item:

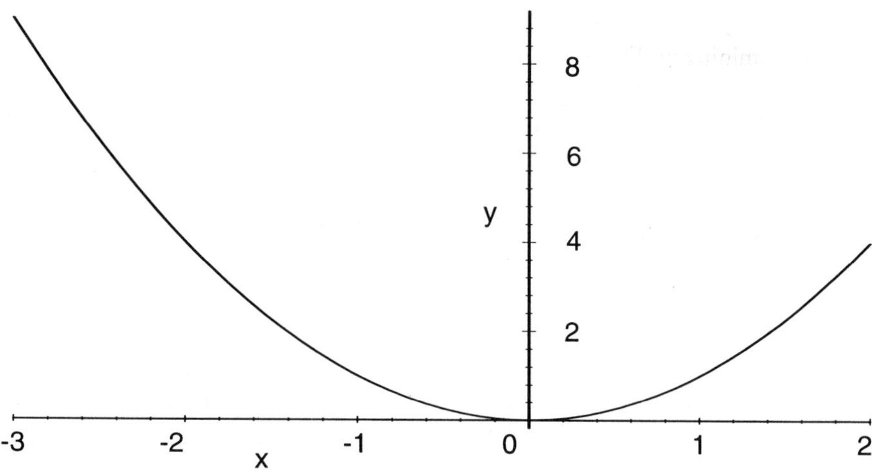

Figure 2.4 Plot of the Function $y = x^2$

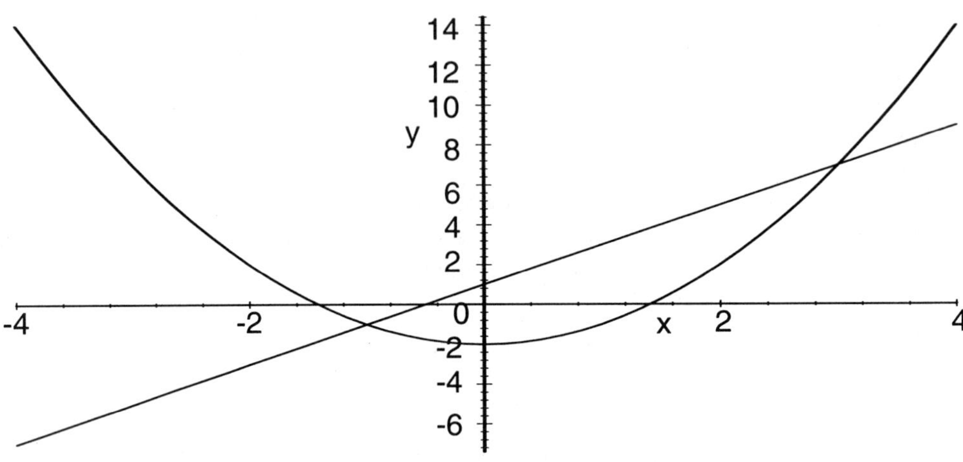

Figure 2.5 Two Functions on the Same Graph

the plot command begins with an open bracket parenthesis and ends with a closing parenthesis. Only the parentheses (as shown) can be used. Here is the *plot* command:

> plot(x^2, x = -3 .. 2);

You can plot two functions on the same graph (Figure 2.5). To do so, you enclose both plot expressions in curly braces, separated by a comma. Say you want to see where the straight line, $y = 2x + 1$, cuts the curve, $y = x^2 - 2$. The plot command is:

> plot({ 2*x + 1, x^2-2 }, x = -4 .. 4);

This graph gives a good idea of where the curves intersect. To find the exact values, you need to use the techniques discussed in the next section.

Your Turn:

1. Plot the function $y = 3x + 2$. You use just the right hand side of the equation in the plot command.

   a) What is the shape of this plot? *Answer:* It is the plot of a __________

   b) Where does the plot cross the x axis? *Answer:* The plot crosses the x axis at x = __________.

2. Plot the function $y = -2x^2 + 6x + 3$. Where does this curve cross the x axis? *Answer:* The plot crosses the x axis at __________ and __________.

3. Plot the function $y = 6x^3 - 10x^2 - 38x + 42$.

a) How many times does this curve cross the *x* axis? *Answer:* The plot crosses the *x* axis __________ times.

b) What are the *x* values there? (These are the solutions to $6x^3 - 10x^2 - 38x - 42 = 0$.) *Answer:* The plot crosses the *x* axis at __________, __________, and __________.

c) One of these values is not a whole number. Express this number exactly. There is more than one way to get the answer. You could use the *solve* command (described next), or you could divide $6x^3 - 10x^2 - 38x + 42$ by $(x-a)$ and $(x-b)$, where a and b are the two integer solutions you found by plotting. In this case, you would use the *divide* command. The form of the divide command for this problem is *divide(6x³ -10x²-38x+42, (x-a)(x-b), `q`);*. The quotient will be put into the variable q. Solving the equation $q = 0$ will give the exact answer for the third root. You could do the division by paper and pencil, since Maple has helped you find the integral solutions!

Answer: The nonintegral root is _________. I used the command ____________________ in my solution.

Solving Equations

Maple's *solve* command is used to find the exact solution to an equation or a set of equations. This may not be possible, and in this case, the *fsolve* command can be used to get a solution to as many decimal places as desired. Let's apply the *solve* command to the problem whose graphical solution is shown in Fig. 2.5. It is written very much like the plot command for two functions.

Example 2-5: Solve the set of equations $y = 2x + 1$ and $y = x^2 - 2$.

```
> solve( { y = 2*x + 1, y = x^2-2 }, {x, y});
                {y = -1, x = -1}, {y = 7, x = 3}
```

Observe that the equations are separated by a comma and they are enclosed in curly brackets. The variables in the problem are also enclosed in curly brackets. The two pairs of curly brackets are separated by a comma and the whole solve command is enclosed in regular brackets.

If there is more than one unknown in a problem, you need a set of equations, one for each unknown. In this case, you use Maple's *set* notation in the *solve* command:

```
> solve( { equation1, equation2, etc. }, { unknown1, unknown2, etc. } );
```

Example 2-6: Solving a Set of Linear Equations

Solve equations such as

$$3x - 2y + 4z = 22$$
$$x - 3y - 2z = -22 \qquad (2\text{-}7)$$
$$4x + 7y + z = 1$$

or the set specified by eq1, eq2, and eq3, for x, y and z.

Solution: Give each equation a name, then use the solve command:

> eq1 := 3*x-2*y+z=-9; eq2 := x+2*y-z=5; eq3 := 2*x-y+3*z=-10;

>solve({ eq1, eq2, eq3 }, {x, y, z});

```
{ x = -1, y = 2, z = -2}
```

Note the special form of the *solve* command. The set of equations and the set of variables have been enclosed in curly brackets, separated from each other with commas. There is also a comma between each item in the set.

Your turn: Solve the set of equations numbered 2-7, above, for x, y, and z

Answer: x=__________, y=__________, z=__________

Example 2-7: The Equation $x^2 - x - 1 = 0$

1. Solve the equation for x. *Answer:* ____________________

2. In Example 2-3, problem 3, the Golden Section was introduced as the number $\frac{\sqrt{5}+1}{2}$.

   In problem 5 of the same set, you were given the number $\sqrt{1+\sqrt{1+\sqrt{1+\sqrt{1+\ldots}}}}$. Call this number τ. Then τ^2 must equal $1 + \sqrt{1+\sqrt{1+\sqrt{1+\sqrt{1+\ldots}}}}$, as you can see by squaring the continued radical once. The second term is just τ once more, so we can write the equation for τ as

$$\tau^2 = 1 + \tau, \textit{ or} \qquad (2\text{-}8)$$
$$\tau^2 - \tau - 1 = 0$$

   Solve this equation for τ using Maple's *solve* command. (You can call the variable t instead of *tau* if you prefer.)

   Answer: Maple gives the solutions ____________________ and ____________________.

   The positive solution is the (same/different) ______________, compared to $\frac{\sqrt{5}+1}{2}$.

3. Here is a type of problem that has interested puzzle solvers for a very long time. A student with 22 cents to spend wants to buy 22 pages of paper. He goes to a store with three types of paper. Plain white (flimsy) sheets are 1/2 a cent each, the better quality white paper costs two cents each and colored bond costs three cents each. He would like to buy some of each kind and use all his money. How many of each can he buy?

   There are three unknowns and only two equations in this problem. If you use the *solve* command, Maple will give you a solution, but it will be in terms of one of the variables. You must find the solution which gives positive whole numbers as answers. Try some substitutions for *z* in the answer set and you will quickly find the answer. Another way to solve the equations is to use the *isolve* command:

   > a := isolve({x + y+z=22, 1/2*x+2*y+3*z = 22 });

   The parameters to this command are just the two equations, separated by a comma and enclosed in curly braces. Maple's answer will be in terms of a variable, _N1, which can be any integer. You can find the value of _N1 which will satisfy the problem by examining the three terms in the answer. Use the *substitute* command to verify your choice.

   Answer: The student can buy ______ flimsies, _______ white and ______ colored pieces of bond paper.

Paper and Pencil Exercises

PP2–1

Add these fractions mentally. Leave the result as an improper fraction.

(a) $\frac{3}{2}+\frac{2}{3}$ *Answer:* ____________________

(b) $\frac{1}{2}+\frac{1}{4}$ *Answer:* ____________________

(c) $\frac{1}{2}+\frac{1}{4}+\frac{1}{8}$ *Answer:* ____________________

PP2–2

Find the lowest common denominator for $\frac{7}{30}+\frac{2}{45}$. *Answer:* ____________________

PP2–3

What is the length of the hypotenuse *H*?

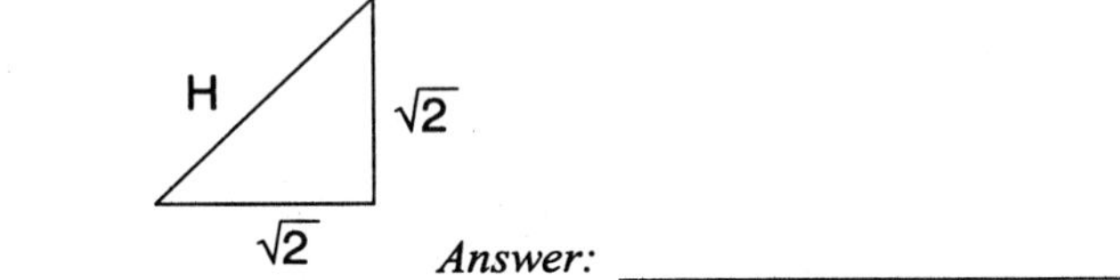

Answer: ______________________

PP2–4

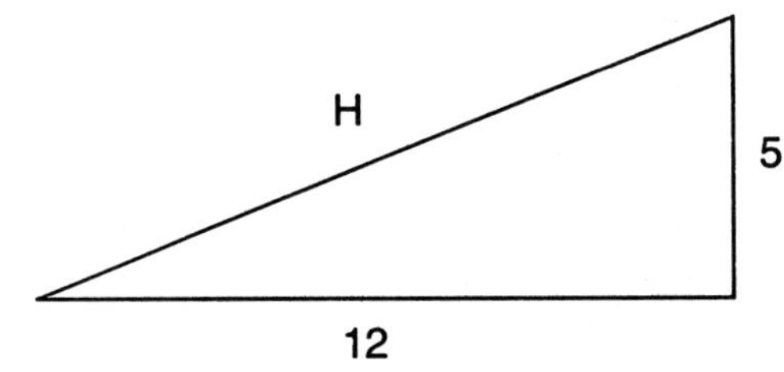

What is the length of the hypotenuse *H*?

Answer: ______________________

PP2–5

Find all the factors of

(a) 45 *Answer:* ______________________

(b) 55 *Answer:* ______________________

(c) 32 *Answer:* ______________________

(d) 1,890 *Answer:* ______________________

(e) 36 *Answer:* ______________________

PP2–6

Mentally compute the following roots. Simplify the roots as far as possible, yet keep the answer exact.

(a) $\sqrt{9} + 2\sqrt{16}$ *Answer:* ______________________

(b) $\sqrt{2} + \sqrt{8}$ *Answer:* ______________________

(c) $\sqrt{4} + \sqrt{9}$ *Answer:* ____________

(d) $\sqrt{9} + \sqrt{16}$ *Answer:* ____________

(e) $\sqrt{3} + \sqrt{27}$ *Answer:* ____________

PP2–7

Estimate the square root mentally, stating which consecutive integers the root lies between.

(a) $\sqrt{2}$ *Answer:* ____________

(b) $\sqrt{5}$ *Answer:* ____________

(c) $\sqrt{10}$ *Answer:* ____________

(d) $\sqrt{26}$ *Answer:* ____________

(e) $\sqrt{37}$ *Answer:* ____________

PP2–8

Factor these expressions.

(a) $x^2 + 2xy + y^2$ *Answer:* ____________

(b) $9 + 12 + 4$ *Answer:* ____________

(c) $x^2 - y^2$ *Answer:* ____________

(d) $100 - 1$ *Answer:* ____________

(e) $ab - ac$ *Answer:* ____________

PP2–9

Solve these equations by paper and pencil. Solve as many as you can mentally.

(a) $x - 3 = 5$ *Answer:* ____________

(b) $mx + b = 4$, for x *Answer:* ____________

(c) $x^2 - 9 = 0$ *Answer:* ____________

(d) $10x + 2 = 7x + 17$ *Answer:* ____________

(e) $3\frac{1}{x} = \frac{9}{3}$ *Answer:* ____________

(f) $\dfrac{2x-1}{x} = 1$ *Answer:* ______________

(g) $\dfrac{x}{2} + \dfrac{x}{3} = 5$ *Answer:* ______________

(h) $\dfrac{1}{x-3} - 2\dfrac{1}{x+3} = 0$ *Answer:* ______________

(i) $5\dfrac{x}{x+3} - \dfrac{3}{4} = 4$ *Answer:* ______________

(j) $10x = 9x + 12$ *Answer:* ______________

(k) $10x = -9x + 12$ *Answer:* ______________

(l) $13 - 2x - 3x - 4x = 5x$ *Answer:* ______________

PP2–10

Express this sum as a single fraction by finding the common denominator.

$\dfrac{a}{b} + \dfrac{c}{d}$ *Answer:* ______________

PP2–11

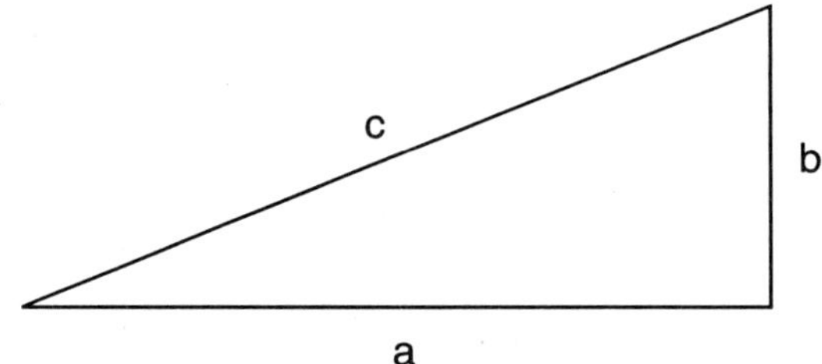

What is the formula for the hypotenuse of this right-angled triangle in terms of the two sides? (Pythagoras' Theorem)

Answer: ______________

PP2–12

(a) Simplify $\sqrt{576 + 4900}$ *Answer:* ______________

(b) Simplify $-\dfrac{1}{2}\dfrac{4x^2y^3-2x^3y^2}{x^2y^2}$ *Answer:* ______________

(c) List the numbers 22/7, π, and $2\sqrt{3}$ in numerical order.

Answer: ______________________________

(d) Solve $\dfrac{6+0.8x}{20+x} = 0.6$ *Answer:* ______________

(e) Solve $x = x_0(1 + \alpha(T - T_0))$ for α. *Answer:* ______________

PP2–13

Simplify $\dfrac{\dfrac{x+y}{x-y} - \dfrac{x-y}{x+y}}{1 + \dfrac{x-y}{x+y}}$ *Answer:* ______________

PP2–14

Solve the quadratic $x^2 - 4x + 4 = 0$ *Answer:* ______________

PP2–15

Simplify $\dfrac{2^{-1/2}}{(\sqrt{3})^{-3}} \cdot \dfrac{\sqrt{2}}{3}$ *Answer:* ______________

Maple Lab

ML2–1

Use Maple to add the following fractions. Leave the answers in fractional form.

(a) 1/2 + 1/3 + 1/4 *Answer:* ______________

(b) 1/1 + 1/1! + 1/2! + 1/3! +1/4! *Answer:* ______________

(c) 3 7/10 + 3 1/7 *Answer:* ______________

(d) $\dfrac{1}{1 + \dfrac{1}{2 + \dfrac{1}{3}}}$ *Answer:* ______________

(e) $(6^{-1} + 7^{-1} + 8^{-1})^{-1}$ *Answer:* ______________

ML2–2

Reduce the following to a 10-figure decimal approximation.

(a) 23/35 + 35/23 *Answer:* ____________________

(b) $\sqrt{2\sqrt{3\sqrt{4\sqrt{5}}}}$ *Answer:* ____________________

(c) Pi^2/exp(1) (These are Maple quantities!) *Answer:* ____________________

(d) Evaluate $G\dfrac{m_1 m_2}{r^2}$, $G = 6.67259\times10^{11}$, $m_1{=}75$, $m_2{=}5.9742\times10^{24}$, $r{=}6.37783\times10^{6}$

Answer: ____

(e) $R_H = 2\dfrac{\pi^2 me^4}{h^3 c}$, $R_H = 109{,}500\ \text{cm}^{-1}$, $c = 2.99792458 \times 10^{10}$, cm/s, $m = 9.1093897 \times 10^{-27}$ kg, $h = 6.6266755 \times 10^{-34}$ erg·s, $e = 4.69 \times 10^{-10}$ esu. Using the above data and the equation, check the value of the Rydberg constant. R_H is an important constant in atomic physics.

Answer: ____

ML2–3

A powerful mathematical technique is called *iteration*. It means doing something again and again, so it is really a fancy word for repetition. We will illustrate the method by using it to find the square root of a number.

Let $x = 1.5$ be an estimate for the square root of two. The formula $\dfrac{1}{2}\left(x + \dfrac{2}{x}\right)$ is a better estimate than x is. Type in the commands

> **x := 1.5; i := 0;** (x is an estimate for the square root of two and i is a counter, initially zero.)

> **x := 1/2*(x + 2/x); i := i + 1;** (A better estimate for the square root each time this command is done.)

Re-execute the last command. In Windows, just place the cursor on the command line and press *Return*. The new estimate is computed and the counter shows how many times the command has been done. Compare these estimates to the square root of two computed by the command

> **evalf(sqrt(2));**

Show your results after you execute the command once, twice, and three times.

> **evalf(sqrt(2));**

Show your results after you execute the command once, twice, and three times.

(a) First time: i = 1, *Answer:* x = ______________, x–sqrt(2) = ______________

(b) Second time: i = 2, *Answer:* x = ______________, x–sqrt(2) = ______________

(c) Third time: i = 3, *Answer:* x = ______________, x–sqrt(2) = ______________

(d) Compare the third result to sqrt(2), evaluated to 10 decimal places. How close are they?

Answer: ____

ML2–4

Edit the formula for finding the approximation to the square root. It becomes

> **x := 1.5; i := 0;** (x is an estimate for the square root and *i* is a counter, initially zero.)

> **x := 1/2*(x + 3/x);** (Make sure *x* has a numerical value before executing this command!)

This time, compare the results to the square root of three. How many iterations does it take to approximate $\sqrt{3}$ to 10 decimal places?

Answer: ____

ML2–5

Try the process again, setting

> **x := 1/2*(x + 10/x);** (Make sure *x* has a numerical value before executing this command!)

How many iterations does it take to get $\sqrt{10}$ to 10 decimal places?

Answer: ____

This problem shows you that Maple makes repetitive calculations easy. You must set up the problem, but then you let Maple take over and do the work.

The techniques of editing commands and re-executing them are real time savers! Learning these techniques early can help you become an efficient Maple user.

ML2–6

Another useful technique to learn is that of solving equations by plotting. This problem asks you to solve

> **eq := x^3 -7*x^2 + x = -44;**

$$eq := x^3-7x^2+x = -44$$

by plotting.

Step 1. Rewrite the equation, collecting all the non-zero terms on the left-hand side.

> **e26 := lhs(eq)-rhs(eq);**

$$e26 := x^3-7x^2+x+44$$

Note that we have used the commands *lhs* and *rhs* to avoid making any errors in copying. Maple does the copying for you, and turns the equation into an expression! You need an expression because you can't plot an equation. Now use the *plot* command:

> **plot(e26, x = -1 .. 10);**

There appear to be three places where the plot crosses the *x* axis. These are the roots of the equation.

(a) Estimate the roots. *Answer:* ____________________

(b) Use Maple's *solve* command and state the solution.

Answer: ____

Note: The *plot* command allows you to find solutions that *solve* can't. Here, both methods work. You can always narrow the plot range to "home in on" a particular solution to get better accuracy.

ML2–7 Estimation

A triangle is determined by giving the coordinates of its three vertices. The triangle for this problem has vertices A (0, 0), B (5, 0), and C (5, 1). You can plot the triangle by listing its coordinates. Maple will "join the dots" if you give the *plot* command a list of the coordinates in order. Starting at point A, which is at the origin, the coordinates are (0,0), (5, 0), (5, 1). Think of going from coordinate point to coordinate point. You will arrive at the point $x = 5$, $y = 1$ by a triangular path. To get back to the origin, you need to go to (0, 0) once more. In Maple Release 3 (V3), the format for this list is simply [0,0,5,0,5,1,0,0]. You don't need to type the round brackets (and), just list the points, separated by commas, inside *square brackets*. Try the command

> **plot([0,0,5,0,5,1,0,0]);**

In Maple Release 4 (V4), you need to keep all the points bracketed. In V4, the command will be

```
> plot( [[ 0,0],[5,0],[5,1],[0,0]]);
```

If you are working in Windows, choose **Projection, Constrained** in the Plot Menu in either V3 or V4. This draws both axes to the same scale so the figure is not distorted. Use a protractor to measure the angle BAC.

Answer: ______________________________

You can measure the angle on the screen or you can print the plot. This has given you a method of measuring an angle without using trigonometry at all!

ML2–8

Another way of estimating the angle is to use the formula for arc length, $s = r\theta$. If you have printed out the triangle to scale, swing an arc, with center at the origin and with radius 5 (inches or centimeters, whatever scale you choose), from the base to the hypotenuse. You can see that the arc length is almost the same as the altitude of the triangle. Use a scale that makes the arc length one unit (5 in = 1 unit, or 5 cm = 1 unit). Estimate the angle using the arc length formula with $s = 1$.

(a) Convert your answer to degrees. *Answer:* ______________

(b) Here is a tricky question in estimation: is the value in (a) too high or too low?

Answer: ______________________________

The next problem asks you to build more right-angled triangles. The base of each is the hypotenuse of the previous triangle in the sequence. You will be able to calculate lengths exactly, but you will have to estimate the angles. Later on, when you have studied more trigonometry, you will know how to calculate the angles. The purpose of this problem, and the next, is to show you that you can "get close" to the answer even if you don't know all the formulas.

ML2–9

Call the hypotenuse AC of triangle ABC in ML2–7 h_1. Construct another triangle with base h_1. Make the altitude 1, just as it is in triangle ABC. Call the hypotenuse of this triangle h_2. Construct another triangle with altitude one, using h_2 as the base. Call the hypotenuse of this triangle h_3.

(a) Solve for h_3 exactly, using Maple.

> ; (Write the Maple command here) *Answer:* ______________

(b) Write the decimal approximation: *Answer:* ______________

(c) Using Maple, estimate the number of triangles you would need for h_n to point nearly straight up. Estimate the angle BAC in triangle ABC by drawing it to scale, or by approximating the angle as = 1/5 radian, using the arc length formula.

Answer: ______________________________

(d) What would the length of the hypotenuse in (c) be?

Answer: ______________________________

(e) As you add more and more triangles on top of each other, when will the hypotenuse h_n be a whole number again? What will this length be?

Answer: ______________________________

(f) What is happening to the size of the angle as we keep on adding triangles on top of one another? Justify your result by examining the relation $\theta = s/r$ as r gets larger and larger.

Answer: ______________________________

(g) What is the length of the hypotenuse of the 42nd triangle? With this many triangles, the spiral of triangles will make a full rotation.

Answer: ______________________________

To verify this last result, you need to apply trigonometry. You will learn the trigonometric solution to this problem in the upcoming labs.

CHAPTER 3

The Three Basic Trigonometric Functions

Objectives for This Chapter

1. Measure lengths using the number scale
2. Extend the measurement process to two dimensions using the rectangular coordinate system
3. Define the trigonometric functions of a general angle
4. Determine which quadrant an angle is in and determine the algebraic sign of the basic trigonometric functions
5. Solve right triangles for angles and sides
6. Apply the trigonometry of right-angled triangles in practical applications
7. Use Maple to solve problems featuring right-angled triangles

Maple Functions Used in This Chapter

sin(x), cos(x), tan(x)	The names of the basic trigonometric functions in Maple.
plot(sin(x), x = 0 .. 2*Pi)	An example of the *plot* command.
evalf(Pi)	Calculate π as an approximate decimal.
evalf(subs(k=150, k*cos(Pi/4))	Find the decimal approximation after substituting a value.
subs(k = 150, k * cos(Pi/4))	Substitute 150 for k in $k\cos(\pi/4)$.

Theory

The basis of trigonometry is the right-angled triangle. We will make the length of the hypotenuse of this triangle 1 unit. Then we will place the bottom corner of the triangle, where the hypotenuse meets the base, at the origin of a rectangular coordinate system. The base has length x, the height (altitude) of the triangle is y, and the angle at the origin is θ. The length of the hypotenuse is H.

The lengths of the three sides of the triangle are related by the Pythagorean theorem, which states that H, x, and y are related as follows:

$$H^2 = x^2 + y^2$$

Drawing the triangle inside a circle, as shown in Figure 3.1, will become even more useful when we consider angles between 0° and 360° later on. Imagine the hypotenuse lying initially on the x axis, so that an angle of 0° is formed. The height, y, is 0. Then think of the hypotenuse rotating around the origin in a counterclockwise direction. It reaches the position shown in Figure 3.1, then continues to rotate into quadrants Q2, Q3, and Q4. The line, y, is always drawn from the point where the hypotenuse meets the circle down to the x axis, so a right-angled triangle is always formed, no matter how large the angle θ.

For the time being, we will focus on a right-angled triangle in the first quadrant, Q1, where the angle θ is between 0° and 90°.

Definitions of Sine, Cosine, and Tangent

The sides of the right triangle are x, y, and H. We can think of comparing one side to another by taking the ratio of the two chosen sides. How many ways can you make up a ratio using the sides of the triangle? It is not difficult to write down all possible ratios. They are y/H, x/H, y/x, H/y, H/x, and x/y. We give these ratios the names sine, cosine, tangent, cosecant, secant, and cotangent, and use the names in a shortened form.

$$\sin(\theta) = \frac{y}{H}$$

$$\cos(\theta) = \frac{x}{H}$$

$$\tan(\theta) = \frac{y}{x}$$

You can see from Figure 3.1 that neither the base or the altitude can exceed the hypotenuse, so $\sin(\theta)$ and $\cos(\theta)$ are always less than or equal to 1. The tangent is the ratio of the altitude to the base. When the angle θ is near 90°, the height, y, is nearly 1, and the base, x, is nearly

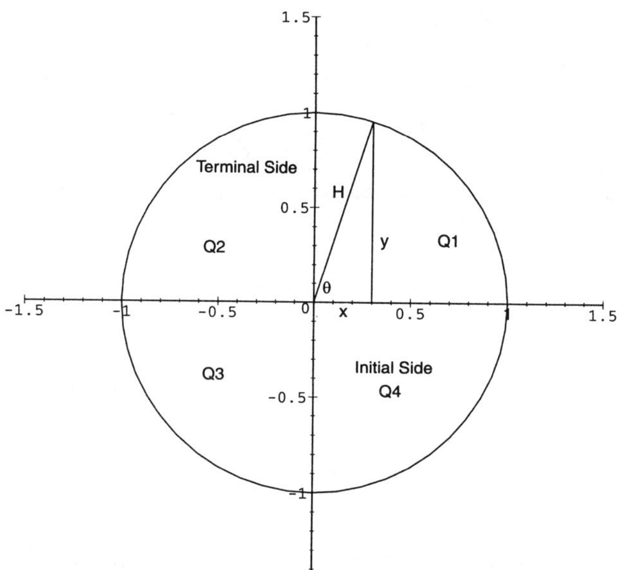

Figure 3.1 The Right Triangle in a Unit Circle Using a Rectangular Coordinate System

0. This situation will make the ratio y/x very large, so the tangent can be any number. It can be 0, less than 1, 1, or greater than 1.

Plotting Basic Trigonometric Functions

If we could animate Figure 3.1 so that the terminal side slowly rotated around the origin, we would see the height, y, change. Beginning with the terminal side lying on the x axis, imagine it rotating counterclockwise. At each instant the height (line y) is drawn and the sine curve is plotted for all values of the angle from 0 up to that point. The diagram is shown in Figure 3.2.

Since $\sin(\theta) = y/H$, and $H = 1$, the line, y, is numerically equal to $\sin(\theta)$. Observe that the x axis is labeled θ because we will plot the points (x, y) as $(\theta, \sin(\theta))$. Look again at Figure 3.2. The angle is measured, and its sine is calculated. Then the height, $y = \sin(\theta)$, is constructed at the value θ on the horizontal axis. This point is $(\theta, \sin(\theta))$. As the line, H, revolves, this diagram traces out the sine curve. The solid curve represents the sine function up to angle θ. The rest of the sine curve, which will be produced as H revolves further, is drawn in with a dotted line. Study this diagram carefully. With practice, you will see the relationship between a rotating line (the hypotenuse of a triangle of ever-increasing angle) and the sine curve.

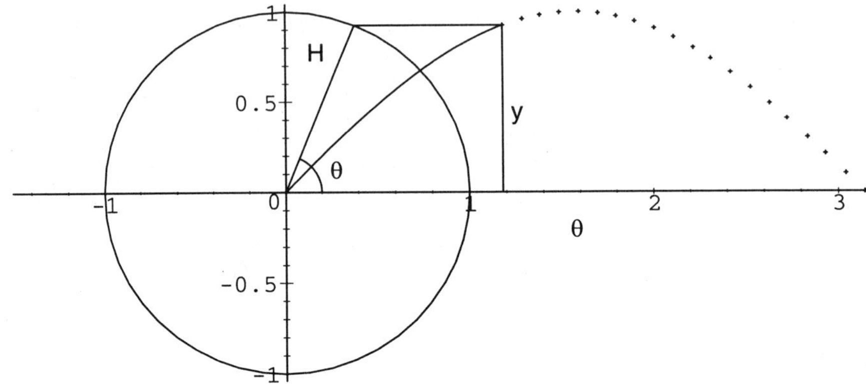

Figure 3.2 The Sine Curve and the Unit Circle

Figure 3.3 shows the diagram for an angle θ in quadrant Q2. When θ enters quadrants Q3 and Q4, sin(θ) will be negative. The full sine curve will be formed when θ goes from 0 to 2π radians.

To plot the trigonometric functions sin(θ), cos(θ), and tan(θ), use the standard Maple *plot* command. The command takes two parameters. The first parameter is the expression to be plotted. You must not use an equation as the plot expression. If you want to plot the equation y = sin(θ), you type in only the *expression,* sin(θ). A comma separates this first parameter from the range expression, which is the second parameter. A range expression has the form, θ = 0 .. Pi, which says that the independent variable θ goes from 0 to π. First, you specify

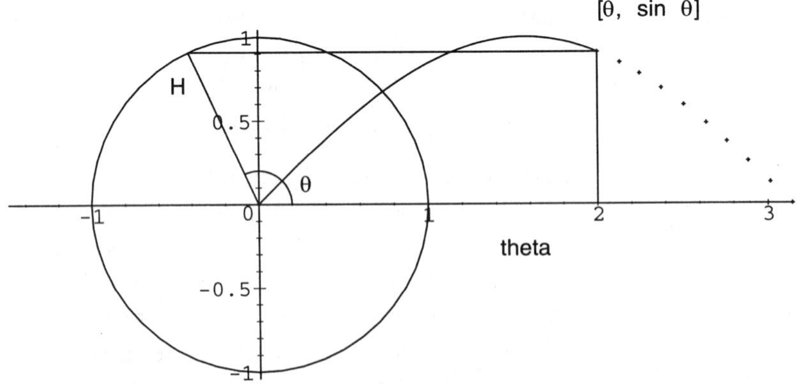

Figure 3.3 The Sine Curve and the Unit Circle for a Larger Angle θ

the independent variable, which must be the same as the only variable in the first parameter of the *plot* command. Then you state the starting-point value, add two periods (with no spaces between them), and finally state the endpoint value. The starting-point value must be less than the endpoint value. When all the necessary components have been assembled, the plot command will be:

> plot(sin(theta), theta = 0 .. 2*Pi);

You must type in the variable θ as *theta* because Maple does not have a facility for automatically inserting letters from the Greek alphabet. If you type in *theta* as part of an expression, Maple will respond with θ wherever *theta* was used. Note an essential difference: an input of *pi* to Maple will be interpreted as the Greek letter pi and will cause Maple to output π, but this symbol π has *no value*. However, if you type in *Pi*, Maple will still output π, but it will now have a value: the ratio of the circumference of a circle to its diameter (see Figure 3.4).

Next we show the commands for plotting cos(θ) and tan(θ), and their plots (see Figures 3.5 and 3.6).

> plot(cos(theta), theta = 0 .. 2*Pi);

> plot(tan(theta), theta = 0 .. 2*Pi, -3 .. 3, title = `Graph of the Tangent Function`, discont=true);

In the case of the tangent function, a more complicated *plot* command is required to help Maple produce a nice-looking graph. If you experiment with the simple plot command:

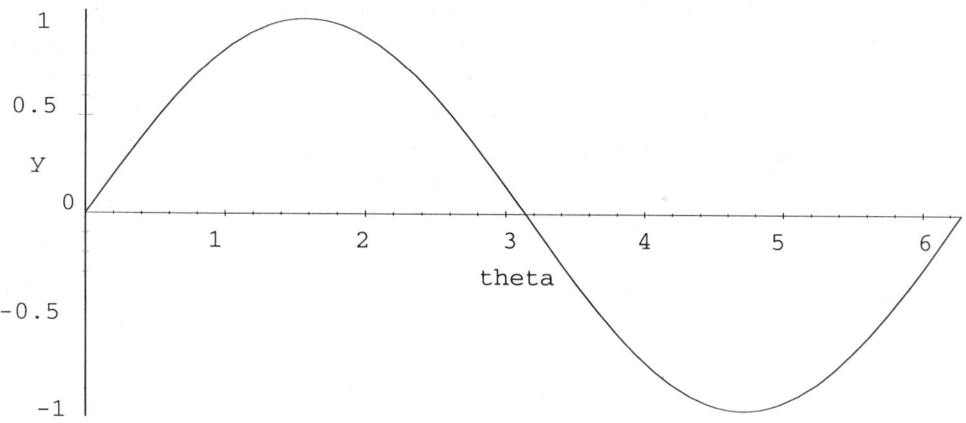

Figure 3.4 The Sine Curve y = sin(θ)

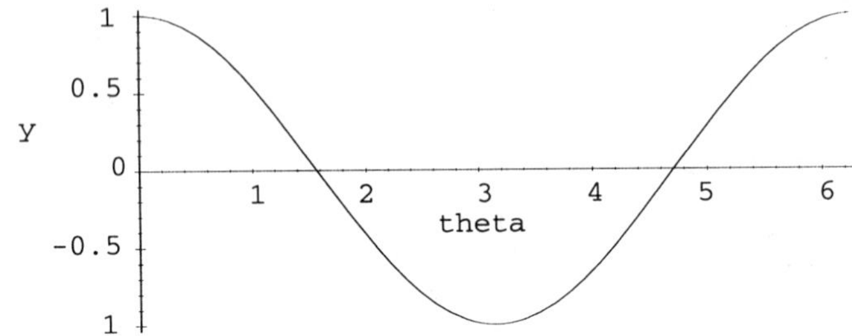

Figure 3.5 Graph of the Cosine Function $y = \cos(\theta)$

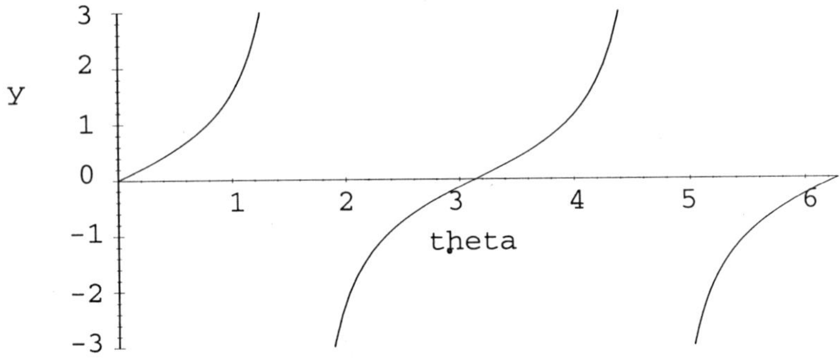

Figure 3.6 Graph of the Tangent Function $y = \tan(\theta)$

> plot(tan(x), x= 0 .. 2*Pi);

you will find that Maple produces an awful-looking plot because the function goes off to $+/-\infty$ at $y = \pi/2$ and $3\pi/2$. Since Maple plots its graphs by evaluating the function at a number of points and drawing a smooth curve between them, it needs to be told that there are breaks in the curve of $\tan(x)$. A section of the *plot* command, *discont* = *true*, performs this function. Maple will automatically scale the y axis as well. In the case of $\tan(x)$, which diverges to $+/-\infty$ in the plot region, we need to set a range for the y axis as well. Finally, we have given the graph a title, using *title* = *`Graph of the Tangent Function`*, where back quotes have been used as a type of brackets around the words we want to appear in the title of the graph. Each of these parameters to the *plot* command is separated by commas. Study the commands carefully so that you will be able to form correct *plot* commands yourself. If you do get a *syntax error* message when you issue a plot command, look for missing commas between parameters, missing double dots in the range specification, and missing parentheses.

There are many things you can learn from Maple's plot. When you issue the *plot* command, the output is placed in a separate plot window in V3 and in V4, it appears in the worksheet

Table 3.1

Plotting the Trigonometric Functions in Degrees

Trig Function	Quadrants Where Pos. +	Quadrants Where Neg. -
sin(θ)	Q1, Q2	Q3, Q4
cos(θ)	Q1, Q4	Q2, Q3
tan(θ)	Q1, Q3	Q2, Q4

right after the *plot* command.[1] You can use the mouse to point to any place on the plot area, and the coordinates will be displayed when you click the left mouse button. By adjusting the position of the mouse pointer after the initial try, you can get fairly close to any point at which you are aiming. For example, if you want to find the value of sin(3), you would put the pointer on the curve, guessing at the x coordinate. Once you click the left mouse button, the x coordinate value will be displayed. Then you can move the pointer to get even closer. Once you are pointing to the curve and the x value is as close to 3 as you can get, you can read the y value as well. The accuracy of this method is no better than two figures, but you can always use Maple's numerical capabilities if you want further accuracy.

A glance at the graphs of the three basic trigonometric functions will allow you to read off the algebraic sign for the function at any given angle. If you want to know the sign of tan(2.7), look at the graph for tan(θ) at θ = 2.7 radians. You see that the plot is below the θ axis, so the value of tan(θ) is negative there. To determine the sign when you do not have the plot of the function in front of you, sketch the "triangle in a circle" for the angle in question. Label the sides of the triangle x, y, and H. Use Table 3.1 and the definition of the trigonometric function to determine the algebraic sign of the result.

Example 3-1: Plotting the Trigonometric Functions in Degrees

To plot sin, cos, and tan in degrees rather than radians, simply issue these plot commands:

> plot(sin(Pi*x/180), x = 0 .. 360); plot(cos(Pi*x/180), x = 0 .. 360); plot(tan(Pi*x/180), x = 0 .. 360);

If x is in degrees, Pi*x/180 is the corresponding angle in radians. Since Maple requires radians in its trigonometric (trig) functions, these commands convert degrees to radians within the trig function while keeping the axis measure in degrees.

1. In V4, you can choose to have the plot appear in a separate window by choosing Options, Plot Display, Window. It is recommended that you use a separate window for your plots.

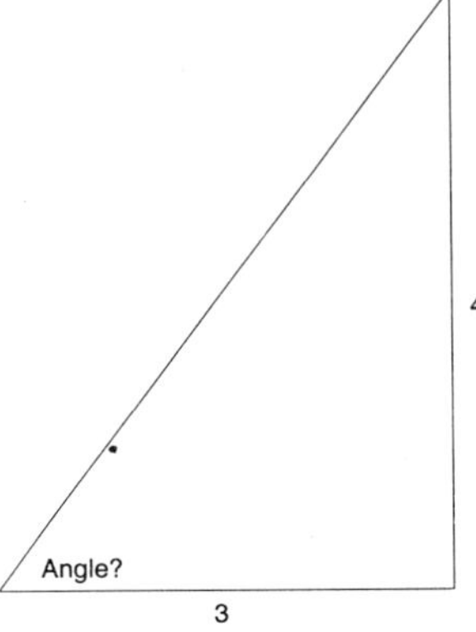

Figure 3.7 Solution of a Right Triangle

Example 3-2: Solution of a Right Triangle

Someone walks three blocks east and four blocks north. The end point of the trip is at an angle to the starting point. What is the sine of this angle? The cosine? The tangent?

Solution. The diagram shows the triangle with the dimensions required by the problem. The base of the triangle has length 3, and the altitude is of length 4. From the dimensions given, the tangent function can be used, since

$$\tan(x) = \frac{y}{x} = \frac{3}{4}$$

Using the Pythagorean theorem, the hypotenuse, $H^2 = 3^2 + 4^2 = 25$; $H = 5$. Therefore, $\sin(x) = y/H = 4/5$ and $\cos(x) = y/H = 3/5$ (see Figure 3.7).

Example 3-3

A surveyor who likes to avoid radian measure tries to measure the height of a building by taking two angle measurements and a length measurement along the base (see Figure 3.8). Here are his readings: angle a = 26.91°; angle b = 43.01°; length k = 150 m. What is the height A?

Solution. This problem can be solved by applying the definitions of the three basic trigonometric functions in a step-by-step manner once you realize that the solution will be assisted by drawing a perpendicular to the hypotenuse, as shown in the diagram. Use the fact that the sum of the angles in a triangle is 180° to find angles $a1$ and $b1$.

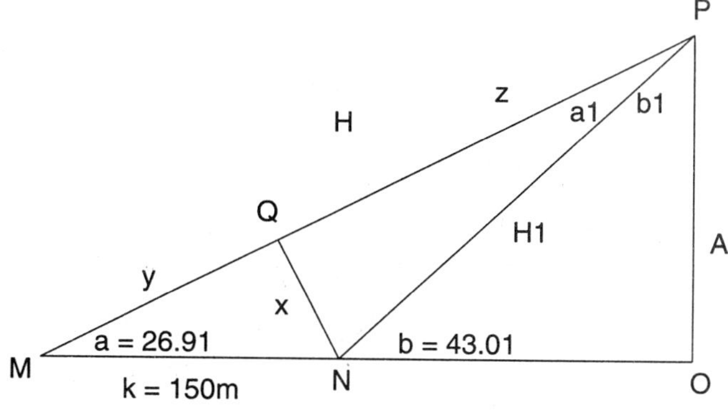

Figure 3.8 Finding the Height of a Building from Angle Measurements

$$b1 = 90 - b = 90 - 26.91 = 46.99^\circ$$

$$a1 = 90 - a - 46.99 = 16.1^\circ$$

$$y = k\cos(a) = 150\cos(26.91) = 133.8 \text{ m}$$

$$x = k\sin(a) = 150\cos(26.91) = 67.98 \text{ m}$$

$$z = \frac{x}{\tan(a1)} = \frac{67.89}{\tan(16.1)} = 235.2$$

$$H1 = y + z = 133.8 + 235.2 = 369 \text{ m}$$

$$A = H1\sin(a) = 369 \text{ x } \sin(29.61) = 167 \text{ m}$$

The height of the building is, therefore, 167 m. You could continue this problem and solve for every length and angle in the diagram just by using the basic definitions.

We have used every one of the basic trigonometric functions to find the solution. The key was drawing the "construction line" to make up the smaller right-angled triangles. This technique can be used quite often to solve problems, because it often reduces a complex problem to a number of simpler problems, which can be solved in turn.

How could the problem be done using Maple? You could type in every computation in the list and have Maple evaluate the answer. You would need to convert the angles to radian measure and you would need to use Maple's *evalf* function to reduce the answers to decimal form. Alternatively, you could analyze the problem using algebraic symbols. Look how the formula for the height can be expressed by working backward from the answer and using $a1 = (90 - a) - (90 - b) = b - a$.

$$A = H1\ \sin(a)$$

$$A = (y + z)\sin(a)$$

$$A = \left(k\cos(a) + \frac{x}{\tan(a1)}\right)\sin(a)$$

$$A = \left(k\cos(a) + \frac{k\sin(a)}{\tan(b{-}a)}\right)\sin(a)$$

Now that we have developed the formula, we can define a Maple expression for A and evaluate it.

> A := (k*cos(a)+ k*sin(a)/tan(b-a))*sin(a);

$$A := \left(k\cos(a) - \frac{k\sin(a)}{\tan{-}b + a}\right)\sin(a)$$

> evalf(subs(k=150, a=26.91*Pi/180, b=43.01*Pi/180, A));

```
166.9891797
```

We typed in the formula, and then used the *subs* command to supply values for k, a, and $a1$. We enclosed the substitute command with an *evalf* command to convert the answer to decimal form. The advantage of the second approach is that we have derived a general formula based on the diagram for the problem. Now we can solve any similar problem by substituting in the new values. Note the actions that needed to be taken:

1. Attempt to reduce the problem to a set of simpler problems, perhaps by drawing in extra lines.
2. Solve the problem in a step-by-step fashion. Use the simple trigonometric formulas to find additional lengths and angles that will be required later.
3. When you have the solution, work backward to express the solution in terms of the given quantities, thus developing an algebraic formula for the solution.
4. Save the result in a Maple worksheet so that you can use your work whenever you need to. You have a *model* that you can use to solve any problem of the same type, simply by substituting in the values appropriate to the new problem.

Paper and Pencil Exercises

PP3–1

Evaluate the trigonometric values for the given angles in radians.

a) $\sin(0)$ ________ b) $\cos(0)$ ________ c) $\tan(0)$ ________

d) $\sin(\pi/2)$ ________ e) $\cos(\pi/2)$ ________ f) $\tan(\pi/5)$ ________

g) $\tan(\pi/4)$ ________ h) $\tan(-\pi/8)$ ________ i) $\sin(1)$ ________

j) $\cos(1)$ ________ k) $\tan(1)$ ________ l) $\sin(2)$ ________

PP3–2

Evaluate the trigonometric values for the given angles in degrees.

a) $\sin(30)$ ________ b) $\cos(30)$ ________ c) $\tan(30)$ ________

d) $\sin(\pi/2)$ ________ e) $\cos(60)$ ________ f) $\tan(45)$ ________

g) $\tan(70)$ ________ h) $\tan(99.5)$ ________ i) $\sin(5)$ ________

j) $\cos(5)$ ________ k) $\tan(5)$ ________ l) $\sin(60)$ ________

PP3–3

Answer these questions about Question PP3–2.

a) Part (d) of question PP3–2 asks you to evaluate $\sin(\pi/2)$. It is unusual that angles in degrees are expressed in terms of π, but it is quite admissible. What is the value of $\pi/2$, expressed to 3 decimal places?

Answer: ____

b) Compare the results for question PP3–2, parts (i), (j), and (k). What can you say about the three values?

Answer: ____

PP3–4

The voltage of an electrical outlet can be described by the equation $V = 117\sin(377t)$, where 117 is a voltage measured in volts, t is measured in seconds, and the angle, $377t$, is in radian measure.

a) What is the voltage at t=1.39 milliseconds (ms)? (1 ms = 0.001 s)

Answer: ____

b) What is the voltage at t = 6.94 ms?

Answer: ____

c) Compare answers (a) and (b). Why are they nearly the same? Convert the two angles to degrees. The two angles are __________ and __________. Sketch the sine curve and mark these angles on your sketch. What is the relationship between these angles?

Answer: ____

PP3–5

A vector is a directed length. Its length is r, and its angle with respect to some reference line is θ. The horizontal component (x component) of a vector is given by the formula $x = r\cos(\theta)$ and its vertical component is given by the formula $y = r\sin(\theta)$.

a) A vector is 12 units long at an angle of 30 degrees. Its y component is __________ units.

b) A vector is 45 units long at an angle of $\pi/6$ radians. Its x component is __________ units.

PP3–6

Fill in the table without using either Maple or your calculator. Instead, use the Pythagorean theorem. For each angle, sketch a right-angled triangle from which to work. (An equilateral triangle has 60° angles. Bisecting one of these angles produces a right-angled triangle with one 30° angle and one 60° angle.)

Table 3.2

Diagram for Problem PP3–6

θ degrees	**θ radians**	**sin(θ)**	**cos(θ)**	**tan(θ)**
30°				
45°				
60°				
90°				

PP3–7

Find sin(θ), cos(θ), and tan(θ) for the triangles in Figure 3.9.

(a) $\sin\theta =$ __________ $\cos\theta =$ __________ $\tan\theta =$ __________

(b) $\sin\theta =$ __________ $\cos\theta =$ __________ $\tan\theta =$ __________

(c) $\sin\theta =$ __________ $\cos\theta =$ __________ $\tan\theta =$ __________

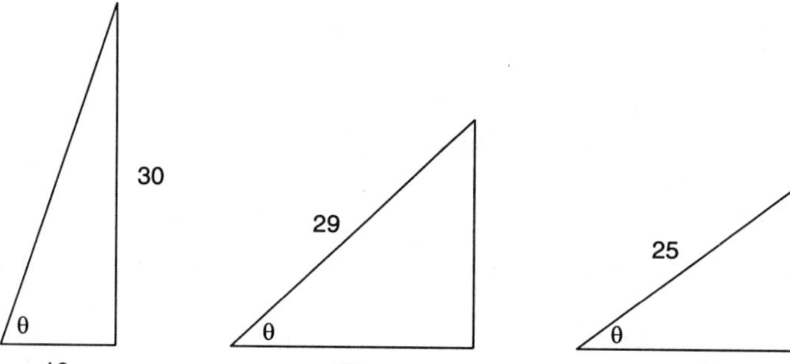

Figure 3.9 Diagram for Problem PP3.7

PP3-8

Fill in the missing values. Assume that all unknown angles are less than 90°.

(a) $\sin(\theta) = 1/2$ $\cos(\theta) =$ ________ $\tan(\theta) =$ ________

(b) $\sin(\theta) =$ ________ $\cos(\theta) = 2/3$ $\tan(\theta) =$ ________

(c) $\sin(\theta) =$ ________ $\cos(\theta) =$ ________ $\tan(\theta) = 5$

Hint: Construct the triangle from the given information. Once you have the triangle, it's easy to read off the values of the other trig function.

Maple Lab

ML3-1

(a) Does Maple work in degrees or radians? *Answer:* ________________

(b) Write the Maple command to find the sine of 37 degrees.

Answer: ____

(c) What is the sine of 1,996 radians? Use Maple to find out. First type

> **sin(1996);**

What is the result? *Answer:* ________________

Why do you get this result?

Answer: ____

____

The way to compute a numerical value for the sine of 1,996 radians is to issue the command:

Write the Maple command here: ______________________________

and the value of sin(1,996), to 10 decimal places, is:

Answer: ______________________________

(d) Sketch the graph of y = sin(x), where x goes from –2*Pi to 2*Pi.

Write the Maple command here: ______________________________

(e) Plot the graph of y = 120*sin(377*t), where t goes from 0 to 1/30th of a second.

Write the Maple command here: ______________________________

(f) Sketch the graph of y = sin(x) + cos(x), where x goes from 0 to 2*Pi. Does it still look like a sine wave?

Describe: ______________________________

(g) Plot the graph of y = 3*tan(x+Pi/4), x = –Pi .. 2*Pi, using the plot options given here. What was the effect of adding Pi/4 to x?

Answer: ______________________________

(h) A pilot in an aircraft flying at 33,000 feet sees a runway ahead. The navigator measures the angle from the airplane to the beginning of the runway as 2.5° below the horizontal. How far is the plane from the runway?

Convert the angle to radians: *Answer:* ______________

Sketch a diagram for the problem:

Answer: ______________________________

Express the unknown length in terms of the known length and a trig function:

Answer: ______________________________

Write the Maple command that solves the problem:

>;______________________________

What happens to the distance if you take half the angle, and then half again?

Answer: ______________________________

ML3–2: Continuation of Example 3-3

While holidaying in Paris in the summer, a surveyor gazes admiringly at the Eiffel Tower from a vantage point across the Seine. She estimates that the top of the tower is 30° above the horizontal. Desiring a closer look, she walks 300 paces (1 pace = 2.5 feet) across the Pont d'Iéna, going directly toward the tower. Now, the top of the tower is 45° above the horizontal. From these measurements, what is the height in feet of the Eiffel Tower?

*Answer:*______________________________ft.

Next summer, the surveyor visits Toronto to take in a Blue Jays game at the Sky Dome. As she drives down John St., she notices she is heading directly for the CN Tower, the highest free-standing tower in the world. She sights to the top of the tower while waiting at a stop light and estimates that the top of the tower is 25° above the horizontal. At the next stop light she rechecks her odometer and finds she has driven 0.3 km from the last light, still going directly toward the tower. Here, the angle appears to be 32°. Estimate the height of the CN Tower, in meters (m).

*Answer:*______________________________m.

Which structure, the Eiffel Tower or the CN Tower, is taller, and by how much?

Answer: The __________________________ is _______________ ft. taller.

ML3–3: A Table of Trigonometric Functions

In the days before the invention of pocket calculators, students bought little books called "trig tables," in which the trig functions were tabulated for every 10th of a degree. You could estimate the value of a trig function to the nearest 100th of a degree by a laborious process called interpolation. Here is a command to tell Maple to make up such a table for the sine function.

> **for i from 0 to 45 do** (The *for* command sets up a repetition for degrees from 0 to 45)

> **for j1 from 0 to .9 by .1 do** (This *for* command sets up a repetition, this one for 10ths of a deg.)

> **printf(`%a\t`, evalf(sin(Pi/180*(i+j1)), 3)); od;** (Calculate the sine of the angle.)

> **printf(`\n`); od;** (Print a new line after every 10 values [i.e., for every degree].)

This is a multiline command. The details are not important here, but notice that the sine function appears on the third line, where the degree value, (*i*), is added to the "10ths of a degree" variable, ($j1$). It is then converted to radians, and the sine is evaluated. Lines 1 and 2 set up the repetition, and line 4 prints a new line after every 10 numbers are printed out.

Type in these commands and observe the output. If you have a word processor available, copy the output and paste it into a text file. The entries in the table should line up fairly well. The first entries are small numbers, which Maple reports in "e-notation."[2]

(a) How would you change the command to have Maple create a table of values for the tangent function?

>; Write the Maple command here ______________________________

(b) Check Maple's output against that from your scientific calculator. Label the rows and columns of the output to show what angle is being computed. What is the largest angle in the table?

Answer: ______________________________

(c) The table only goes a little past 45°. (This was done to save paper.) Take any angle and compute sin(θ) and sin(90° – θ). Do this a few times and record one of the representative examples. What do you conclude about sin(θ) and sin(90° – θ)?

θ = ______________________________

sin(θ) = ______________________________

sin(90° – θ) = ______________________________

Conclusion: ______________________________

(d) Can you think of some instances where it is useful to have printed tables of some mathematical functions? Maple could be used to generate these tables.

Answer: ______________________________

---

2. Maple uses a variation of scientific notation, common in computer science, where numbers are written as a number between 0.1 and 1 multiplied by a power of 10. Thus, the number 0.0123 can be written 0.123e–1.

CHAPTER 4

Reciprocal Trigonometric Functions of an Acute Angle

Objectives for This Chapter

1. Define the reciprocal trigonometric functions
2. Plot the reciprocal trigonometric functions
3. Determine these properties of the reciprocal trigonometric functions: domain, range, zeros, period, and symmetry properties
4. Solve trigonometric identities involving the six trigonometric functions

Maple Commands Used in This Chapter

simplify	Attempt to find a simpler form of an algebraic or trigonometric expression.
sec(x), csc(x), cot(x)	Maple's names for the reciprocal trigonometric functions.
evalf(expr)	Convert *expr* to a decimal number.
plot(sec(x), x = 1 .. 1.52)	Plot the secant function for a range of values.
plot(expr, discont=true)	Plot a curve with a discontinuity (eliminates the vertical line).
plot(expr, title=`Title of Plot`)	Give the plot a title, in this case, "Title of Plot."
rhs	Pick out the right-hand side of an equation.
lhs	Pick out the left-hand side of an equation.

Definition of the Reciprocal Trigonometric Functions

You learned the definitions of the three basic trigonometric functions, sin(x), cos(x), and tan(x), in Chapter 3. The reciprocals of these three basic functions have been given names as well. You know from algebra that the reciprocal of a number is found by dividing that number into 1. That is, the reciprocal of x is $1/x$, and the reciprocal of $1/x$ is x.

The following are the definitions of the reciprocal trigonometric functions:

The Reciprocal Trigonometric Functions

$$\cos ecant(\theta) = \csc(\theta) = \frac{1}{\sin(\theta)} = \frac{H}{y} \tag{4-1}$$

$$\sec ant(\theta) = \sec(\theta) = \frac{1}{\cos(\theta)} = \frac{H}{x} \tag{4-2}$$

$$\cot angent(\theta) = \cot(\theta) = \frac{1}{\tan(\theta)} = \frac{x}{y} \tag{4-3}$$

In these formulas, H is the hypotenuse of the triangle, x is the base, and y is the altitude (see Figure 3.1).

The reciprocal trigonometric functions are not used as frequently as sin, cos, and tan, but they do appear from time to time. Observe the following points concerning these functions:

1. All three can be easily expressed in terms of sin(θ), cos(θ), and tan(θ). This is the reason why they are not used very frequently.
2. The Maple names for these functions are the same as the abbreviated name shown in Equations 4-1 through 4-3: csc(θ), sec(θ), and cot(θ).
3. Always use functional notation. This means putting parentheses around the angle.
4. The range of sin(θ) and cos(θ) is from –1 to 1, but csc(θ), sec(θ), and cot(θ) can grow very large, both negatively ($-\infty$), and positively ($+\infty$). Finding the range of these functions will be left as an exercise. (See Paper and Pencil Exercises for this chapter.)

Graphs of the Reciprocal Trigonometric Functions

The graphs of the reciprocal functions show that they are indeed functions, that is, they pass the vertical line test, which states that for any given value of the angle, a vertical line intersects the curve only once if it intersects it at all. The basic trigonometric functions, sin, cos, and tan, are defined for all angles. The reciprocal trigonometric functions are undefined

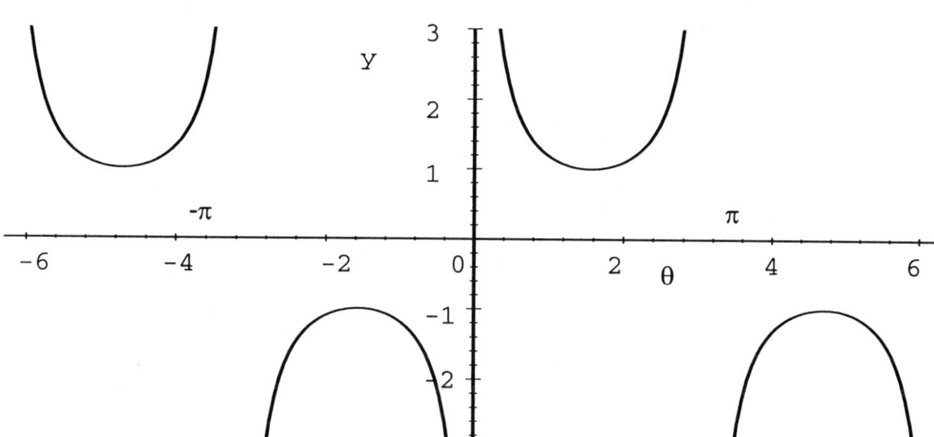

Figure 4.1 The Cosecant Function

at exactly those places where the corresponding basic function is 0. Consider the cosecant function. It is undefined at angles 0, $\pm\pi$, $\pm 2\pi$, $\pm 3\pi$, $\pm 4\pi$.... Near these values of the angle, the cosecant function grows very large, either positively or negatively. The slightly unruly behavior of the sec and csc graphs is one reason why these reciprocal trigonometric functions are not as commonly used as the basic trigonometric functions sin and cos. Both the tangent and the cotangent functions become very large at certain angles, yet the tan function remains the most used of the two (see Figure 4.1).

The secant function is undefined at 0, $\pm\pi/2$, $\pm 3\pi/2$, $\pm 5\pi/2$, $\pm 7\pi/2$ ..., while the cotangent function is undefined at angles 0, $\pm\pi$, $\pm 2\pi$, $\pm 3\pi$, $\pm 4\pi$ ..., just like the cosecant function (see Figures 4.2 and 4.3).

Fundamental Trigonometric Identities

Now that you have seen all six basic trigonometric functions, you can investigate the relationships between them. Keep in mind that all these trigonometric functions arise from ratios of the sides of a right-angled triangle. Since the Pythagorean theorem, $x^2 + y^2 = H^2$, is true for any right-angled triangle, you can always find the third side if you know *any* two sides. This result allows you to calculate all the trigonometric functions if you know any one of them. Specifically, if you know the value of any one of $\sin(x)$, $\cos(x)$ $\tan(x)$, $\cot(x)$, $\sec(x)$, or $\csc(x)$ and the quadrant it is in, you can calculate all the others. This raises an interesting question: why do we need six functions when one will do? The answer is that with six functions, it is easier to find a simple expression for common relationships. For instance, in

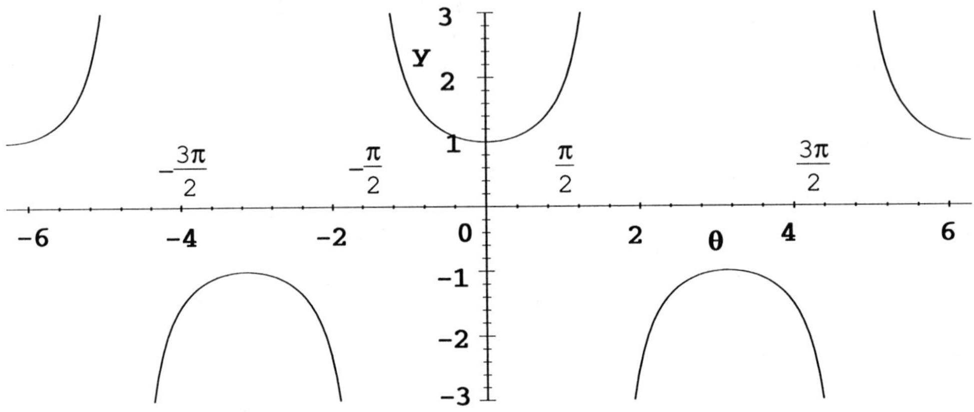

Figure 4.2 The Secant Function

the study of mechanics and electronics, the hypotenuse and the angle at the base of a triangle may be known, and expressions for the base and the height of the triangle may be needed.

If we had only the sine function, the formula for the base (using the Pythagorean theorem) would be (see also Figure 4.4):

$$x = H\sqrt{1 - \sin^2(\theta)}$$

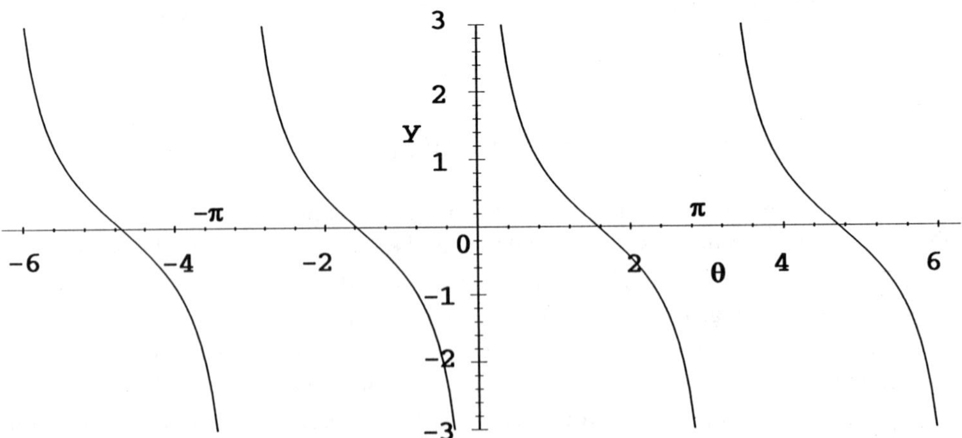

Figure 4.3 The Cotangent Function

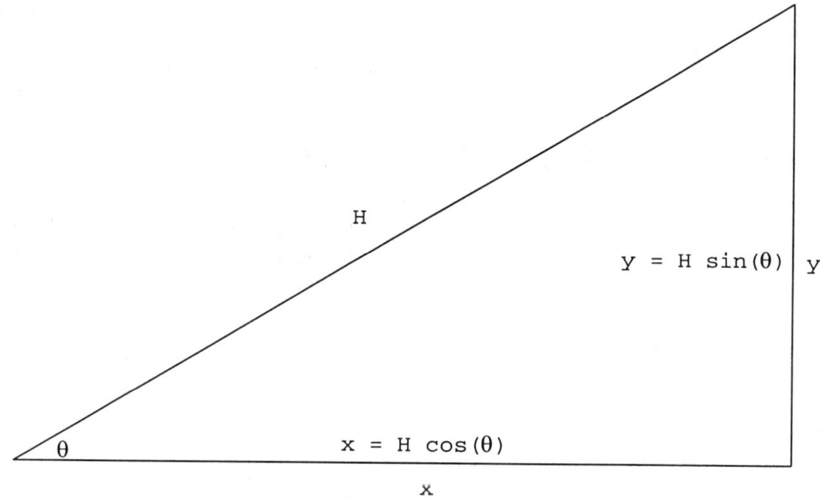

Figure 4.4 Formulas for the Base, x, and the Height, y, in terms of the Hypotenuse, H, and the Angle θ

This formula is more cumbersome because it includes a square root. You would need to use many more keystrokes to calculate the result this way, whether you use a pocket calculator or Maple. It is for this reason that mathematicians have preferred to define six closely related functions. It gives us more flexibility in mathematical expressions.

An *identity* is an equation which remains true for *all* values of the variable contained in it. Here are our first trigonometric identities. They follow immediately from the definitions of csc(θ), sec(θ), and cot(θ).

The Fundamental Trigonometric Identities

$$\sin(\theta)\csc(\theta) = 1, \cos(\theta)\sec(\theta) = 1, \tan(\theta)\cot(\theta) = 1 \quad (4\text{-}5)$$

One way of showing that these relationships are true is to use Maple:

> sin(x)*csc(x); simplify(sin(x)*csc(x));

```
sin(x) 1 csc(x)
```

You see that you needed to use the *simplify* command. If you simply multiply the sin(x) term by the csc(x) term, Maple gives you the same result back again, without change. However, when you tell Maple to *simplify* the expression, you get the result you want.

Obviously, you can derive this result by yourself, without Maple. However, you should use Maple to work out the other two identities just to practice the method. You will use the same technique to prove much more elaborate trigonometric identities later on.

Proving that an identity is true may be difficult if there are many terms in an expression, but proving that a proposed identity is false requires only one counterexample. If you can find some value that causes the two sides of the equation to evaluate to different numbers, you have proven that the equation is not an identity. This can be done by plotting the two sides of the equation on the same graph. You will see an example of this technique in the next section.

Example 4-1

Prove that the equation, $1 + \tan(\theta)^2 = \sec(\theta)^2$, is an identity.

Note that we have written the equation in the style of Maple. The exponent, 2, is written at the end of the expression. On the other hand, you will find it written this way in most textbooks: $1 + \tan^2(\theta) = \sec^2(\theta)$. The meaning is the same: first, you evaluate the trigonometric function; then, you square the result. This is *completely different* from squaring the angle first and then looking up the trig function. Make sure you understand the difference between, say, $\sin(\theta^2)$ and $\sin^2(\theta) = \sin(\theta)^2$. The last two expressions are the same. To refer to them, we say "sine squared theta." To refer to $\sin(\theta^2)$, we say, "sine theta squared." Pick out some values for θ and evaluate both forms using Maple or your pocket calculator. You will see that they are different (unless you are very lucky and manage to solve the equation $\sin(\theta^2) = \sin(\theta)^2$—the solution is $\theta = 0$. Any other value of θ between 0 and 2π will give different values for the right-hand and left-hand sides of the equation).

Now go back to the original problem. You can solve the problem by referring to the defining triangle (see Figure 4.5).

Step 1: Write down the equation you are trying to prove as an identity (Equation 4-6).

$$1 + \tan(\theta)^2 = \sec(\theta)^2 \tag{4-6}$$

Step 2. Use the definition of the trigonometric functions to express them in terms of H, x, and y (Equation 4-7).

$$1 + \frac{y^2}{x^2} = \frac{x^2}{x^2} + \frac{y^2}{x^2} = \frac{x^2 + y^2}{x^2} \tag{4-7}$$

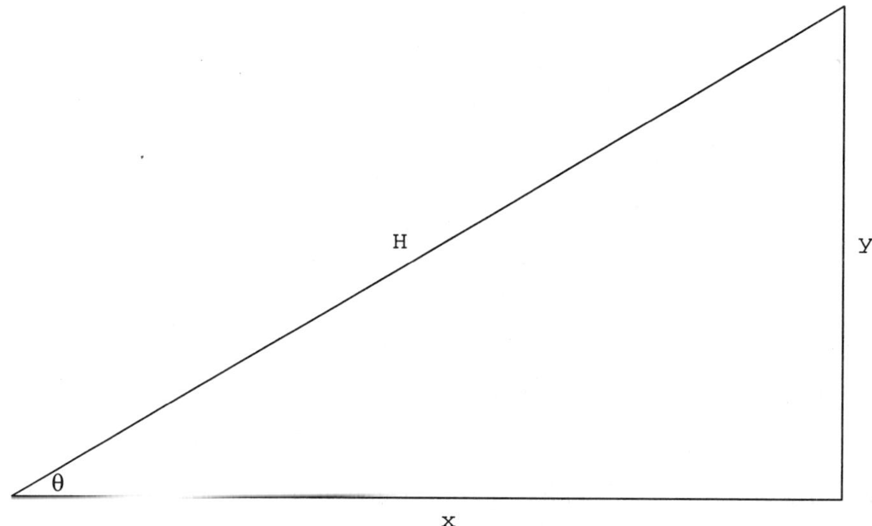

Figure 4.5 Use the Basic Triangle to Help Prove Identities

Step 3: Choose one side of the equation and work with it. In this case, we have chosen the left-hand side (Equation 4-8).

$$\frac{x^2 + y^2}{x^2} = \frac{H^2}{x^2} \tag{4-8}$$

Step 4. Using algebra, simplify the expression.

$$\frac{H^2}{x^2} = \sec(\theta)^2$$

Step 5. Once again using trigonometric definitions, express the simplified result in terms of a trigonometric function.

Example 4-2

Prove that $1 + \cot(\theta)^2 = \csc(\theta)^2$.

This time we will use Maple. The technique is to subtract the right-hand side of the equation from the left-hand side. If the equation is an identity, the subtraction will yield 0.

Step 1: form the Maple statement as follows:

> 1 + cot(theta)^2-csc(theta)^2;

$$1 + \cot(\theta)^2 - \csc(\theta)^2$$

Step 2: simplify the result.

> simplify(1 + cot(theta)^2-csc(theta)^2);

$$0$$

The subtraction does yield 0, so $1 + \cot(\theta)^2 = \csc(\theta)^2$ is an identity.

Your Turn:

(a) Prove that $1 - \csc^2(\theta) = -\cot^2(\theta)$ is an identity.

Answer: ______________________________

(b) Prove that $\sin(\theta)^3 = \sin(\theta) - \cos(\theta)^2 \sin(\theta)$ is likely to be an identity by graphing the two sides of the equation.

Answer: ______________________________

Example 4-3

Prove that $50\dfrac{\cos(x)\csc(x)}{\cot(x)^2} = 49\tan(x) + 1$ is *not* an identity.

To demonstrate that an equation is not an identity, all that is required is to show that the right-hand side does *not* equal the left-hand side for some value (or values) of x. This can be accomplished by plotting both sides of the equation. You simply make some choice for the range of x, like $x = -\pi/2 .. \pi/2$. You make a guess at the range for y, like $-50 .. 50$. If the range you choose the first time doesn't give you a good-looking graph, make another choice. Maple does the hard work of plotting the points and producing the graph.

Step 1. Plot the left-hand and right-hand sides on the same graph (see Figure 4.6).

> plot({ 50*cos(x)*csc(x)/cot(x)^2, 49*tan(x)+1}, x = -Pi/2 .. Pi/2, -50 .. 50);

Step 2. Look carefully at the graph. If you see two lines, the curves for the two sides of the equation are different. Two curves are visible in this plot, so the equation cannot be an identity.

Step 3. To make sure of your result, you can magnify a portion of the graph.

> plot({ 50*cos(x)*csc(x)/cot(x)^2, 49*tan(x)+1}, x = -0.56 .. -0.54);

Here, we have magnified the graph by plotting a small range of values around $x = -0.55$. Try this plot yourself. The result is two clearly separate, almost straight, lines.

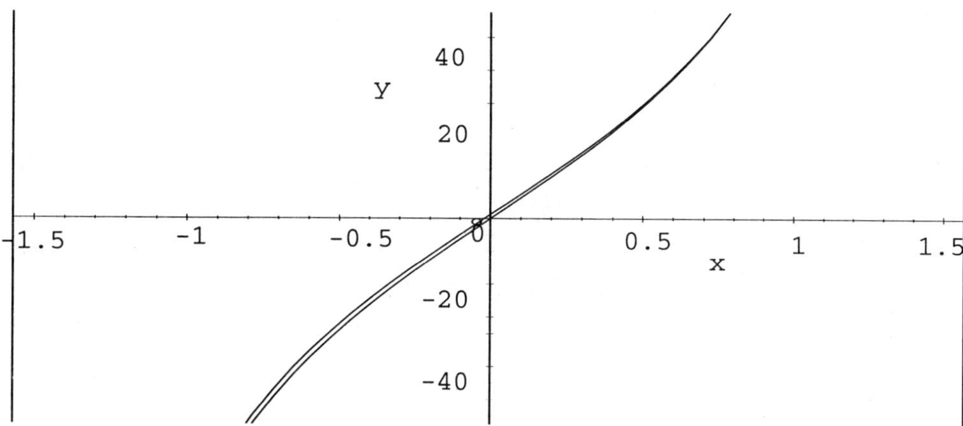

Figure 4.6 Plot of $50\dfrac{\cos(x)\csc(x)}{\cot(x)^2}$ and $49\tan(x) + 1$, Showing That the Two Curves Are Different

Your Turn: Prove that $\sqrt{\dfrac{1+\cos(\theta)}{1-\cos(\theta)}}\sin(\theta) = 1 - \cos(\theta)$ is *not* an identity.

Answer: ____

Example 4-4

Prove that $\tan(y)\csc(y) = -\sec(y)$ is *not* an identity.

Since we have said that it takes only one value of y where the right-hand side is not equal to the left-hand side to disprove the identity, we can pick a point at random and see if the two sides are equal. Let $y = 1$, then:

```
> evalf(tan(1)*csc(1)); evalf(-sec(1));
                    1.850815718
                   -1.850815718
```

We see that when $y = 1$, the right-hand side is the negative of the left-hand side. This should lead us to suspect that there has been a typographical error and that $\tan(y)\csc(y) = \sec(y)$ might be an identity. This is easily proven using Maple:

```
> simplify( tan(y)*csc(y)-sec(y) );
```

$$0$$

Summary.

1. Use the basic triangle (Figure 4.5) to express trigonometric functions in terms of H, x, and y if you want to simplify an expression containing more than one trigonometric function.
2. Plot the right-hand and left-hand sides of a trigonometric equation. The plot will lead you to the solution, which may be an identity.
3. To confirm that some trigonometric equation is indeed an identity, use Maple. Subtract the right-hand side from the left-hand side and simplify the result. If this is 0, you know you have an identity. Note: Maple cannot reduce every such expression to 0. It may need some help from you!

Your Turn. Prove that $\cot(\theta)\sec(\theta) - \tan(\theta)\cos(\theta) = -\cos(\theta)\cot(\theta)$ is *not* an identity.

Answer: ______________________________

Example 4-5

Graphically solve the trigonometric equation $\csc(x) = \cot(x) + 2$. Trigonometric equations are more complicated than either linear equations or quadratic equations. Even if you need an analytic (exact) solution, you will find it helpful to get a picture of the situation by graphing the problem. An organized approach is required because of the complexity of these types of problems. Here are the steps to follow:

1. Write the equation in Maple, giving it a name. You will refer to this equation, or parts of it, again, so you need to be able to identify the equation by name.

2. Plot both sides of the equation on a single graph and see where the two curves cross each other. These points are the solutions for the given equation. Of course, you must choose a range for the plot. A good first guess for the horizontal range is –Pi .. Pi. Since the reciprocal trig functions do go on to infinity, you will probably need to specify a vertical range as well. You will probably need to refine the plot ranges after observing the first attempt.

3. Adjust the plot range until the point of intersection of the two curves is near the center of the plot.

4. In the graph window, place the mouse pointer at the point of intersection and click the left mouse button. The x, y coordinates of the point are displayed at the lower left portion of the window.

5. If you require more accuracy, narrow the horizontal range around the point of intersection.

6. Use Maple's *solve* command on the original equation to attempt an analytic solution. Maple may not be able to find one. If it can't, you will get another prompt (>), but you won't get a message telling you that it could not find a solution. Be careful! Trigonometric equations may have many solutions, and Maple typically finds only one. That is why graphing the problem is so important. Note: if Maple reports a solution with the letter I in it, Maple has found a solution in the domain of complex numbers (in Maple, $I = \sqrt{-1}$). Solutions of this type should be discarded, since we are looking for *real* solutions. Complex numbers are discussed in Chapter 8, Maple for Algebra.

Solution. (See also Figure 4.7).

Step 1:

> eq1 := csc(x) = cot(x) + 2;

```
eq1 := csc(x) = cot(x) + 2
```

Step 2:

> plot({ rhs(eq1), lhs(eq1) }, x = -Pi .. Pi, -5 .. 5);

Step 3: you may adjust the range parameter if you wish. You can see that $x = 2.1 .. 2.3$ might be a good choice. You would not need to specify the vertical range this time—why? Try plotting with the new horizontal range with and without specifying the vertical range to see the difference.

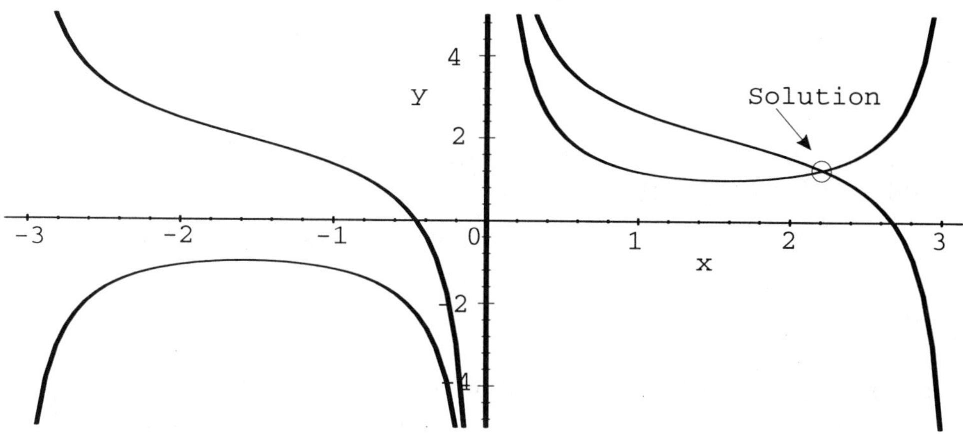

Figure 4.7 Graphical Solution of $\csc(x) = \cot(x) + 2$

Step 4: the approximate solution $x = 2.2$, $y = 1.3$, is read from the bottom-left of the plot window after clicking on the intersection point. *This is the solution to the problem.* It may be approximate, but if you wish, you can increase the accuracy by narrowing the plot range.

Step 5: if we narrow the plot range to $x = 2.2 .. 2.3$, we now find $x = 2.214$, $y = 1.251$. At this magnification, the plot looks like the intersection of two straight lines. We could continue zooming in on the intersection point, if need be, to get even more accuracy.

Step 6: use the Maple *solve* command:

> solve(eq1, x);

Because we gave the equation a name, this step becomes very easy. Of course, we may or may not get a solution, but it sure didn't hurt to try! Maple does find a solution in this case. It is $x = 2$ arctan(2), which evaluates to 2.214 297436. The first four digits found by the graphical method were correct. How would you find the height, y, of the point of intersection? You would substitute $x = 2$ arctan(2) in either csc(x) or cot(x) + 2. Maple gives the result as 1.250. This result is exact, but we will need some additional trigonometric identities in Chapter 10 to prove it.

Your Turn. Solve the equation $\frac{1}{2}[(\tan(x) + \cot(x)) \sin(x) \cos(x)] = \cos(x)$ on the interval $0 \leq x < \pi/2$ by graphing.

Answer: ______________________________

Paper and Pencil Exercises

PP4–1

If $\sec(\theta) = 2\sqrt{2}$ find $\cot(\theta)$. Solve the problem exactly by finding the lengths of the triangle in Figure 4.5. Then find a decimal approximation to the exact solution.

Answers: Decimal approximation: __________ Exact solution: __________

PP4–2

Find all six trigonometric functions for angle θ, given the triangles in Figure 4.8.

(a) $\sin(\theta)$ __________, $\cos(\theta)$ __________ $\tan(\theta)$ __________

$\csc(\theta)$ __________, $\sec(\theta)$ __________, $\cot(\theta)$ __________

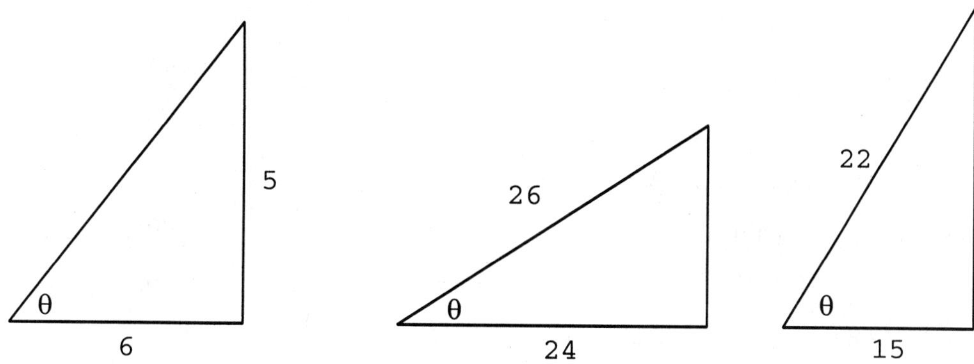

Figure 4.8 Diagram for Problem PP4-2

(b) sin(θ) ________, cos(θ) ________, tan(θ) ________

csc(θ) ________, sec(θ) ________, cot(θ) ________

(c) sin(θ) ________, cos(θ) ________, tan(θ) ________

csc(θ) ________, sec(θ) ________, cot(θ) ________

PP4–3

Find all six trigonometric functions for angle θ, given the lengths of two sides of triangle ABC (see Figure 4.9).

(a) AB = __42__, BC = __40__, AC = ________

Answers: sin(θ) ________, cos(θ) ________, tan(θ) ________

csc(θ) ________, sec(θ) ________, cot(θ) ________

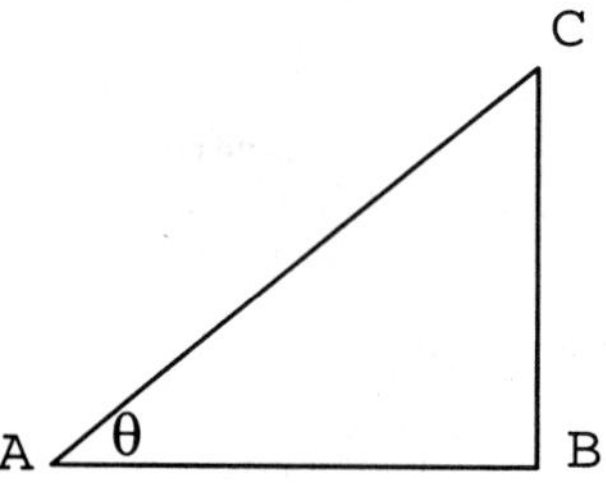

Figure 4.9 Diagram for Problem PP4-3

(b) AB = ___56___, BC = ___33___, AC = ________

Answers: sin(θ) ________, cos(θ) ________, tan(θ) ________

csc(θ) ________, sec(θ) ________, cot(θ) ________

(c) AB = ________, BC = ___24___, AC = ___74___

Answers: sin(θ) ________, cos(θ) ________, tan(θ) ________

csc(θ) ________, sec(θ) ________, cot(θ) ________

(d) AB = ___154___, BC = ________, AC = ___170___

Answers: sin(θ) ________, cos(θ) ________, tan(θ) ________

csc(θ) ________, sec(θ) ________, cot(θ) ________

(e) AB = ________, BC = ___48___, AC = ___290___

Answers: sin(θ) ________, cos(θ) ________, tan(θ) ________

csc(θ) ________, sec(θ) ________, cot(θ) ________

PP4–4

Fill in the missing values. All angles are in the first quadrant.

(a) $\sin(\theta) = 1/2$

Answers: sin(θ) ________, cos(θ) ________, tan(θ) ________

csc(θ) ________, sec(θ) ________, cot(θ) ________

(b) $\cos(\theta) = 1/\sqrt{2}$

Answers: sin(θ) ________, cos(θ) ________, tan(θ) ________

csc(θ) ________, sec(θ) ________, cot(θ) ________

(c) $\tan(\theta) = 5/3$

Answers: sin(θ) ________, cos(θ) ________, tan(θ) ________

csc(θ) ________, sec(θ) ________, cot(θ) ________

(d) $\cot(\theta) = 1.6$

Answer: sin(θ) ________, cos(θ) ________, tan(θ) ________

csc(θ) ________, sec(θ) ________, cot(θ) ________

(e) $\sec(\theta) = 1.5$

Answer: $\sin(\theta)$ ________, $\cos(\theta)$ ________, $\tan(\theta)$ ________

$\csc(\theta)$ ________, $\sec(\theta)$ ________, $\cot(\theta)$ ________

(f) $\csc(\theta) = 4$

Answers: $\sin(\theta)$ ________, $\cos(\theta)$ ________, $\tan(\theta)$ ________

$\csc(\theta)$ ________, $\sec(\theta)$ ________, $\cot(\theta)$ ________

PP4-5

If $\theta = 30°$, evaluate:

(a) $\sin(\theta)\csc(\theta)$ *Answer:* ____________

(b) $\tan(\theta)\cot(\theta)$ *Answer:* ____________

(c) $\cos(\theta)/\sin(\theta) - \cot(\theta)$ *Answer:* ____________

(d) Is the answer dependent on the angle? *Answer:* ____________

PP4-6

Evaluate, for any angle:

(a) $\csc(-x) + \csc(x)$ *Answer:* ____________

(b) $\sec(-x) - \sec(x)$ *Answer:* ____________

(c) $\cot(x) + \cot(-x)$ *Answer:* ____________

PP4-7

Answer these questions about the cosecant function:

(a) The *domain* is *Answer:* ____________

(b) The *range* is *Answer:* ____________

(c) Does the function have any zeros? *Answer:* ____________

PP4-8

Answer these questions about the secant function:

(a) The *domain* is *Answer:* ____________

(b) The *range* is *Answer:* ____________________

(c) Does the function have any zeros? *Answer:* ____________________

PP4–9

Answer these questions about the cotangent function:

(a) The *domain* is *Answer:* ____________________

(b) The *range* is *Answer:* ____________________

(c) Does the function have any zeros? *Answer:* ____________________

PP4–10

Given the triangles (a), (b), and (c) in Figure 4.10, find the values of the inverse trigonometric functions for each.

(a) csc(θ) __________, sec(θ) __________, cot(θ) __________

(b) csc(θ) __________, sec(θ) __________, cot(θ) __________

(c) csc(θ) __________, sec(θ) __________, cot(θ) __________

PP4–11

Fill in the missing values.

(a) csc(θ) ____5____, sec(θ) __________, cot(θ) __________

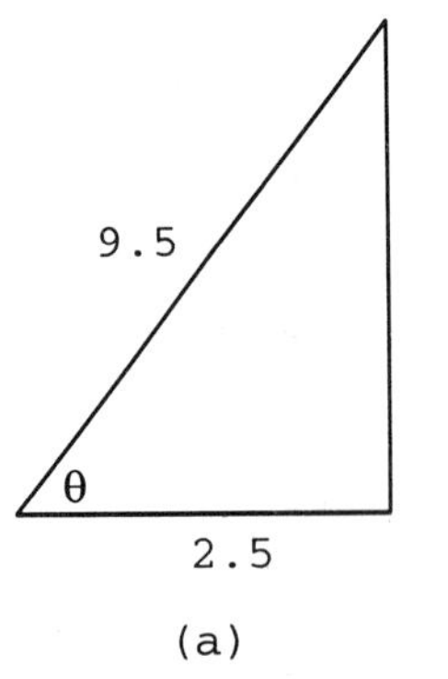

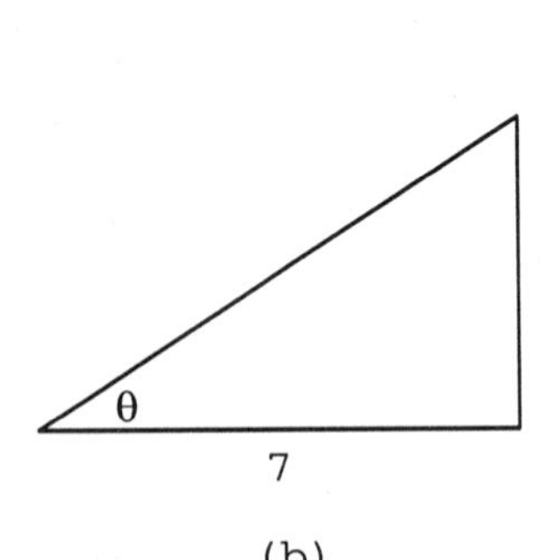

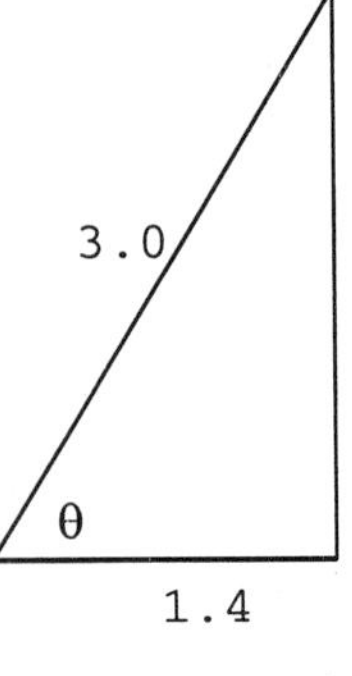

Figure 4.10 Diagram for PP4-10

(b) $\csc(\theta)$ ________, $\sec(\theta)$ ___2___, $\cot(\theta)$ ________

(c) $\csc(\theta)$ ________, $\sec(\theta)$ ________, $\cot(\theta)$ ___1/3___

Maple Lab

ML4–1

One way of "proving" that $\sin^2(\theta) + \cos^2(\theta) = 1$ is to plot the left-hand side of this equation over one full cycle. If the plot is a horizontal straight line at a height equal to 1, then the function is a constant.

> **plot(sin(theta)^2 + cos(theta)^2 , theta = 0 .. 2*Pi);**

Describe the plot:

Answer: ____

ML4–2

Extend the range of the plot in ML4–1 to θ = –2*Pi .. 2*Pi. Describe the result:

Answer: ____

ML4–3

Plot $\cos^2(\theta)-\sin^2(\theta)$ over the range θ = –2*Pi .. 2*Pi. Is this the plot of a constant function? If not, describe its major features like the maximum point(s), minimum point(s), and location of the zeros of the function.

Answer: Maxima at ____

Answer: Minima at ____

Answer: Zeros at ____

ML4–4

Prove that $2\sin^4(x) - 3\sin^2(x) - 1 = \cos^2(x)\,(1 - 2\sin^2(x))$ is *not* an identity. Plot each side of the equation. If the plots are different at just one point, the equation cannot be an identity.

> ; (Write the *plot* command here) ____

Is the equation an identity?

Answer: ____

ML4–5

Another way of showing that an equation is not an identity is to evaluate each side for some random (or selected, if you wish) values of the variable. Evaluate the equation in ML4–4 for some values of x in the interval 0 .. 2π. Some values of x may satisfy the equation. These are the *solutions* to the equation in the specified interval. All it takes is one value of x that makes the two sides different to prove that the equation is not an equality. Find one of these values for x. State the value of x and the different values of the left-hand side (lhs) and right-hand side (rhs) of the equation.

>; (Write the Maple command here) ______________________________

Answer: x value = __________, lhs = __________, rhs = __________

ML4–6

You have learned that $\sin^2(\theta) + \cos^2(\theta) = 1$ is an identity. Justify this fact by computing the left-hand side of the equation for various values of θ. Issue the following commands:

> e46 := evalf(sin(theta)^2 + cos(theta)^2); (Note the difference in the way you write *sin²(θ)* in Maple)

```
e46 := sin(θ)^2 + cos(θ)^2
```

> evalf(subs(theta = 0, e46));

Edit the last line, setting $\theta = 0.1, 0.2, 0.4, 1, 2, 3, 4, 5$, and 6, in turn. Verify that the result is 1 each time.

Are all the values *exactly* 1? Account for any discrepancies.

Answer: ______________________________

ML4–7

Show that $\sin^2(\theta) - \cos^2(\theta)$ is not an identity by repeating ML4–6 with the + sign changed to a – sign. Edit *e46* in this way and report one of the values of θ that proves your case.

Answer: θ = __________, $\sin^2(\theta) - \cos^2(\theta)$ = __________

ML4–8

Maple works in radian measure. Your scientific calculator contains a "switch" to permit you to set it to work in radians or degrees. There are advantages and disadvantages to this approach. Many slips are made because people forget which mode they are working in. Maple always works in radians. The only slip you can make is to forget this fact.

Now that you have been reassured that Maple has closed the door on one possible way of making a slipup, we will reopen the door! You can teach Maple to use degrees as well as radians. You have seen one method already, where the special names *deg_* and *rad_* were defined. The method employed in this problem is to define a new set of functions. We distinguish these new functions from the built-in functions by capitalizing the first letter. Remember that Maple is "case sensitive," which means that it treats capital letters as completely distinct from their lower-case partners. For example, Maple considers the two names, *Cos* and *cos*, as completely different. You can find a fuller explanation of Maple functions in *Maple for Algebra*, Chapter 2, in this series.

Define a new function:

> **Sin := x -> sin(Pi*x/180);**

$$\text{Sin} := x \to \sin\left(\frac{1}{80}\pi x\right)$$

If you have not seen this way of defining a *function* before, please accept it for now. You will gradually come to investigate these functions to show that they do work.

(a) Define Cos, Tan, Sec, Csc, and Cot

> ; (Write the Maple definitions here)

Answers:

Cos: > ____________________

Tan: > ____________________

Sec: > ____________________

Csc: > ____________________

Cot: > ____________________

(b) Type the definition in a new Maple worksheet. Save the worksheet as *tdeg.ms* (Maple V3) or *tdeg.mws* (Maple V4). Whenever you want to work in trigonometry and evaluate trig functions of angles in degrees, load this worksheet and reexecute the commands that define Sin, Cos, and so forth. Continue on with any other work you need to do, working with degrees if you need to. Save your completed worksheet with the *Save As* . . . option in the *File* menu. In this way, your original file, *tdeg.ms* (or *tdeg.mws*), will remain unchanged. An added bonus will be that the "degree" commands will be included in every worksheet you create this way.

Define the commands of ML4–8(a), and save them to the file. Open this file as a Maple worksheet. Scroll down to the bottom of the worksheet and issue the command:

> Sin(45);

What is the output? Explain what happened.

Answers: Output: ____________________

Explanation: ____________________

(c) You can open file *tdeg.mws* (V4) or *tdeg.ms* (V3) and copy the entire worksheet. Next, open another worksheet and paste in the commands from *tdeg.mws*. This is a way of incorporating these commands into any existing worksheet.

Must you reexecute these commands before they will work?

Answer: ____________________

ML4–9

Use the commands you defined in ML4–8 to evaluate the following. Check your answers with a calculator.

(a) Sin(30) ________	Sin(150) ________
(b) Cos(150) ________	Cos(510) ________
(c) Tan(45) ________	Tan(135) ________
(d) Tan(90) ________	Tan(180) ________
(e) Sin(60) ________	Sin(120) ________
(f) Sec(45) ________	Csc(135) ________
(g) Sec(450) ________	Sec(135) ________
(h) Sin(45) ________	Sin(135) ________
(i) Cos(45) ________	Cos(135) ________
(j) Cos(60) ________	Cos(120) ________

ML4–10

Plot $1 + \cot(x)^2$ and $\csc(x)^2$ on the same graph. What does an examination of the graph tell you?

> ; (Write the Maple command here) ____________________

Answers: Plot range chosen: ____________________

Graph shows: ______________________________

ML4–11

(a) What is the Maple command that evaluates (in decimal) the cotangent of 57.3°?

Answer: ______________________________

(b) What is the Maple command that evaluates (in decimal) the cotangent of 1 radian?

Answer: ______________________________

M4-12

(a) What is the result of this command?

> csc(Pi); *Answer:* ______________

(b) Try evaluating csc(π) on your calculator. *Answer:* ______________

(c) Referring to Figure 4.1, explain the problem: *Answer:* ______________

ML4–13

Spot the error:

(a) csc x ; *Answer:* ______________

(b) 4 cot(x) ; *Answer:* ______________

(c) 2 * sec * (x) ; *Answer:* ______________

ML4–14

What is the Maple command to plot the square of the cotangent function in the range –1 to +1 radians?

> ; (Write the Maple command here) ______________________________

ML4–15

Why is it better to use our new function, Csc, when we want to plot the cosecant function in terms of degrees instead of radians? After all, you could issue the command:

> plot(csc(Pi/180*x), x = 0 .. Pi);

Answer: ______________________________

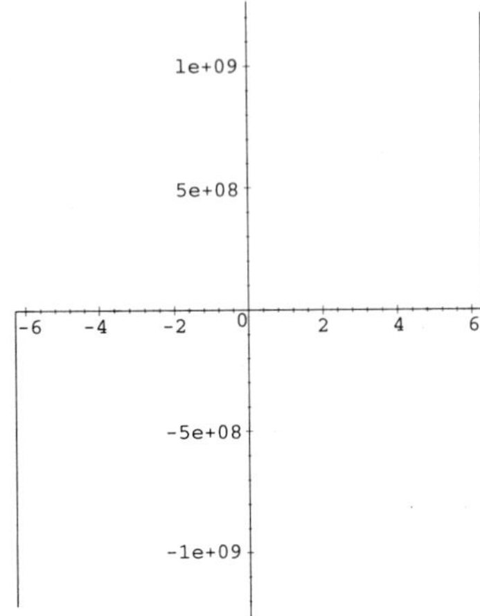

Figure 4.11 A Cosecant Function Plot with No Useful Information (Diagram for ML4–16)

ML4–16

You can get a lot of information by plotting a function. To get the most from your plot, you need to choose horizontal and vertical scales that show the characteristics of the function as clearly as possible. Observe what can go wrong if you issue the following plot command:

> plot(csc(x), x = -2*Pi .. 2*Pi);

See Figure 4.11. What happened is that the cosecant function exploded to $\pm\infty$ at $x = \pi$ and $x = -\pi$. When Maple attempts to plot the function, it encounters some very large positive and negative values, so it scales its graph accordingly. Unfortunately, this results in a nearly unreadable graph. The graph shown in Figure 4.1 was produced by giving Maple a vertical range as well as a horizontal range:

> plot(csc(x), x= -2*Pi .. 2*Pi, -Pi .. Pi, discont = true, title = `Cosecant Function`);

Notice two additional things about this plot command. The use of *discont* = *true* prevents Maple from "joining the dots" as the function makes large jumps near the values $x = -\pi$ and $x = \pi$. You can title a graph using *title* = *`Cosecant Function`*. Notice that you enclose the text in special quotation marks, called *backquotes*. As usual, these different parts of the plot command are separated by commas. Try issuing this command without putting in *discont* = *true* and note the difference. Also, experiment with titling your graphs, which adds a professional touch.

1. Answer the following questions by referring to the graph of the cosecant function shown in Figure 4.1.

(a) What is the *domain* of the cosecant function? (Remember: places where the function is *not defined* cannot be part of the domain!)

Answer: ______________________________

(b) What is the *range* of the cosecant function? (Be careful: for instance, is the value 0 included in the range?)

Answer: ______________________________

(c) What is the *period* of the function? (The period is the shortest horizontal distance over which the function repeats.)

Answer: ______________________________

(d) State the *zeros* of the function, the values of x where $y = 0$ in $y = \csc(x)$.

Answer: ______________________________

(e) Does the function exhibit symmetry? Is the function even or odd? If the function is even, then $\csc(-x) = \csc(x)$ and if it is odd, $\csc(-x) = -\csc(x)$. *Hint:* graph $\csc(x)$ and $\csc(-x)$ on the same plot. You can determine the symmetry of the function by examining the result of the plot.

Answer: ______________________________

2. Answer the following questions by referring to the graph of the secant function shown in Figure 4.2.

(a) What is the *domain* of the secant function? (Remember: places where the function is *not defined* cannot be part of the domain!)

Answer: ______________________________

(b) What is the *range* of the secant function? (Be careful: for instance, is the value 0 included in the range?)

Answer: ______________________________

(c) What is the *period* of the function? (The period is the shortest horizontal distance over which the function repeats.)

Answer: ______________________________

(d) State the *zeros* of the function, the values of x where $y = 0$ in $y = \sec(x.)$

Answer: ______

(e) Does the function exhibit symmetry? Is the function even or odd? If the function is even, then $\csc(-x) = \csc(x)$, and if it is odd, $\sec(-x) = -\sec(x)$. *Hint:* graph $\sec(x)$ and $\sec(-x)$ on the same plot. You can determine the symmetry of the function by examining the result of the plot.

Answer: ______

3. Answer the following questions by referring to the graph of the cotangent function shown in Figure 4.3.

(a) What is the *domain* of the cotangent function? (Remember: places where the function is *not defined* cannot be part of the domain!)

Answer: ______

(b) What is the *range* of the cotangent function? (Be careful: for instance, is the value 0 included in the range?)

Answer: ______

(c) What is the *period* of the function? (The period is the shortest horizontal distance over which the function repeats.)

Answer: ______

(d) State the *zeros* of the function, the values of x where $y = 0$ in $y = \cot(x)$.

Answer: ______

(e) Does the function exhibit symmetry? Is the function even or odd? If the function is even, then $\csc(-x) = \csc(x)$ and if it is odd, $\cot(-x) = -\cot(x)$. *Hint:* graph $\cot(x)$ and $\cot(-x)$ on the same plot. You can determine the symmetry of the function by examining the result of the plot.

Answer: ______

4. In which quadrants are the $\csc(x)$, $\sec(x)$, and $\cot(x)$ functions positive?

Answers: $\csc(x)$ ______

$\sec(x)$ ______

$\cot(x)$ ______

5. Does the function $y = \csc(x)$ have a minimum for values of x *between* 0 and Pi? Stated more precisely, does $y = \csc(x)$ have a minimum for values of x for $0 < x < \pi$?

Answer: ______________________________

6. Does the function $y = \sec(x)$ have a minimum for $0 < x < \pi$?

Answer: ______________________________

7. Does the equation $\sin(x) = \csc(x)$ have any solutions? If so, where?

Answer: ______________________________

Explorations

E4-1

(a) You have seen that the trigonometric functions are strongly interrelated. Use the Pythagorean identities

$$\sin(\theta)^2 + \cos(\theta)^2 = 1$$

$$\sec(\theta)^2 = 1 + \tan(\theta)^2$$

$$\csc(\theta)^2 = 1 + \cot(\theta)^2$$

to express $\cos(\theta)$, $\tan(\theta)$, $\cot(\theta)$, and $\sec(\theta)$ in terms of $\sin(\theta)$.

(b) Express each of the remaining five trigonometric functions in terms of $\tan(\theta)$.

E4-2

In the analysis of a trigonometric problem, formulas may be derived that are not necessarily in simplest form. Simplify each of the following.

(a) $\tan(\theta) + \dfrac{\cos(\theta)}{1 + \sin(\theta)}$

(b) $\cot(\theta) + \dfrac{\sec(\theta)}{1 + \csc(\theta)}$

E4-3

Verify the following identities.

(a) $\dfrac{1 - \sin(\theta)}{\cos(\theta)} = \dfrac{\cos(\theta)}{1 + \sin(\theta)}$

(b) $\dfrac{\cos(\theta)\cot(\theta) - \sin(\theta)\tan(\theta)}{\csc(\theta) - \sec(\theta)} = 1 + \sin(\theta)\cos(\theta)$

E4-4

(a) Express $\dfrac{\sin(\theta)}{\sin(\theta) + \cos(\theta)}$ in terms of $\sec(\theta)$ and $\csc(\theta)$.

(b) Express $\dfrac{\sec(\theta) + \csc(\theta)}{\tan(\theta) + \cot(\theta)}$ in terms of $\sin(\theta)$ and $\cos(\theta)$.

E4-5

Simplify $(\tan(a) + \tan(b))(1 - \cot(a)\cot(b)) + (1 - \tan(a)\tan(b))(\cot(a) + \cot(b))$

CHAPTER

5

Trigonometric Functions of a General Angle

Objectives for This Chapter

1. Express angles as rotations and parts of a rotation
2. Define trigonometric functions of a general angle and relate them to the reference angle
3. Completely solve the right-angled triangle given any two values
4. List cofunction relationships
5. Define and use periodic functions
6. Investigate recurrence relationships

Maple Commands Used in This Chapter

evalf(expr)	Evaluate *expr* as a decimal number.
irem(25, 4)	Divide 25 by 4 and state the remainder, i.e., 1.
isolate(eqn, var)	Rearrange *eqn*, bringing *var* by itself to the *lhs*.
simplify(expr)	Simplify an expression which can contain trigonometric functions.
trunc(number)	Truncate a number to the next nearest integer toward 0.
with(student)	Load the student package to make the *isolate* command available.

Trigonometric Functions of a General Angle

The three basic trigonometric functions, sin, cos, and tan, were defined in Chapter 3 and the inverse functions, sec, csc, and cot, were defined in Chapter 4. The algebraic sign of the basic trig functions in any quadrant was defined in Chapter 3 as well. You can see from the definition of the inverse trigonometric functions that each has the same sign as its related function. That is, cot(x) will have the same sign as tan(x), sec(x) will have the same sign as cos(x), and csc(x) will have the same sign as sin(x). We will now extend these definitions and results to angles greater than 2π radians, or 360°.

Rotations

The terms *initial side* and *terminal side* were introduced in Figure 3.1. They form the two sides of an angle. Here are some more definitions to make the discussion of general angles easier. The *vertex* of an angle is the point where the two sides meet. It is the point (0, 0) in Figure 3.1. Now think of two angles, θ_1 and θ_2. The first angle, θ_1, is constructed by placing its initial side along the x axis with its vertex at the origin. Its terminal side makes an angle of 60°. The diagram is shown in Figure 5.1, part (a).

Figure 5.1, part (b), shows the result of rotating the terminal side through one full rotation and a further 60°. If you placed part (b) on top of part (a), the two angles would lie on top of each other. They would be indistinguishable. The only difference between the two diagrams is the curly line, which indicates that more than a full revolution has been made.

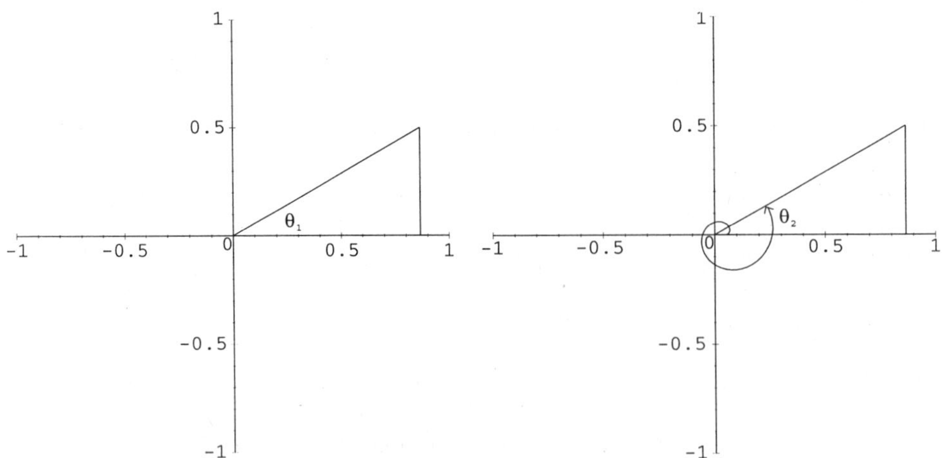

Figure 5.1 An Angle of 60° and an Angle of 420°

We call angles like 60° and 420° ($\pi/6$ radians and $2\pi + \pi/6$ radians) *coterminal angles*. If you rotate the terminal side through any number of full revolutions before stopping, you will create an angle coterminal with one that has been turned for less than a full rotation. Turning the process around, you can always express any angle, like θ_2, no matter how large, as an angle θ_1 less than 360°. These angles are the same in respect to the values of the trigonometric functions. Note the implications carefully. The *position* of a tire that has turned more than one rotation may be the same as the position of one that has turned less than a full rotation, but the tread wear will be quite different.

Consider a car tire that has been placed on a test fixture in a research and development lab. The tire is mounted above a moving belt in order to measure tread wear. The axle is attached to an angle-measuring device, which counts revolutions. After a prolonged test, the tire may have rotated through an angle of 3.0×10^9 degrees. If a mark had been placed on the bottom of the tire when the experiment started, where would it be now? We can divide the angle by a number of full rotations using Maple. To see how many rotations were made, divide by 360°.

```
> Number_of_rotations := trunc(evalf(3e9/360));
```

```
                Number_of_rotations := 8333333
```

The number of degrees in the remaining partial rotation must be:

```
> 360*( 3.0e9/360-Number_of_rotations);
```

```
                          120.
```

Another way of arriving at the same result is to use remaindering. We divide 3×10^9 by 360 and ask Maple for the remainder:

```
> irem( 3000000000, 360);
```

```
                          120
```

We conclude that the mark on the tire is at an angle of 120° to the negative y axis, that is, at an angle of 30° to the x axis. After this calculation, we know that 120° is somehow equivalent to 3 billion degrees! But in what way, exactly? Certainly not in terms of tread wear! A brand-new tire that has rotated through an angle of 120° is still brand new, while a tire that has been on-test for 3 billion degrees has traveled the equivalent of:

```
> evalf(1*ft*(3e9*Pi/180)/(6280*(ft/mi)), 3);
```

```
                        8320. mi
```

In this calculation, we have used the formula $s = r\theta$, converted the angle to radians, and converted feet to miles. We have also used the indeterminates *ft* and *mi* to show the units explicitly. The tire has gone over 8,000 mi. Even though its position may be the same as if

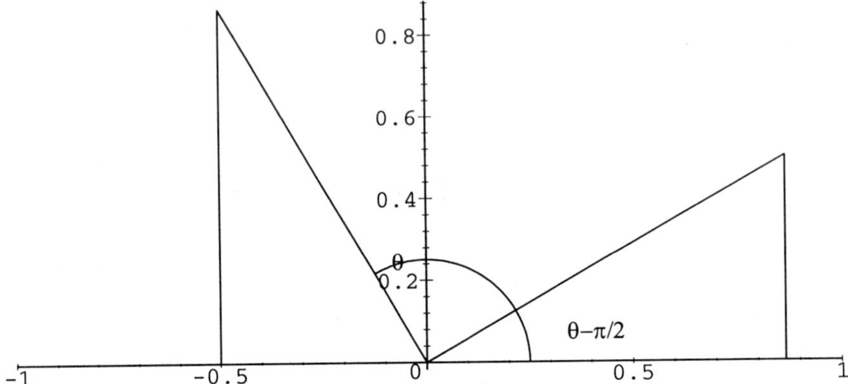

Figure 5.2 The Angles θ and θ-π/2 Are *Not* Equivalent

it had rotated only 120°, a lot else has happened. Nevertheless, sin(120) = sin(3,000,000,000). The trigonometric functions for the two angles are the same.

The angle 120° is called the *reference angle*.[1] As far as the trigonometric functions go, their values are the same whether they are evaluated at 3,000,000,000° or 120°. The reference angle is defined as the angle less than 360° coterminal with the given angle. To find the reference angle, you subtract multiples of 360° from the given angle until a remainder of 360° is obtained.

1. Note that the definition of the *reference angle* given here differs from the standard one found in most textbooks. The reason for changing the definition in this book is that it makes it more useful when using Maple or, for that matter, your scientific calculator. The definition given here preserves the sign of the trigonometric function. The usual definition is that the reference angle is the acute angle θ′ formed by the terminal side of θ and the x axis. You can always calculate this angle given our definition of the reference angle and using the diagram in Figure 5–2(a). Remember always to draw in the vertical line joining the terminal side to the x axis, even if the terminal side lies below that axis.

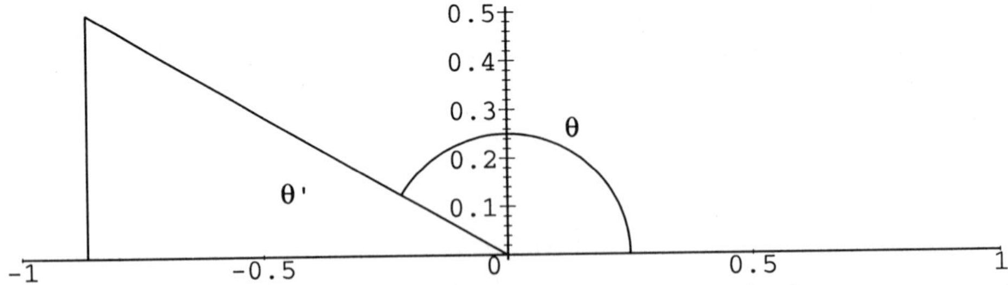

Figure 5.2(a) Definition of *Reference Angle* Found in Most Textbooks

Can this process be taken further? Why not subtract multiples of 90° until an angle less than a right angle is reached? Would the trigonometric functions of these angles be the same as the original? Unfortunately they would not. As Figure 5.2 shows, angles θ and $\theta-\pi/2$ are not equivalent. The diagram shows angles $\theta = 120°$ and $\theta - \pi/2 = 30°$. They are related, as we will see later in Chapter 8, but they are not the same.

The trigonometric functions of a general angle are defined in terms of the equivalent reference angle. This definition has an important consequence: all the trigonometric functions can have their angles increased (or decreased) by multiples of 360° (or 2π radians) without changing their values. If n is an integer, then the following relationships hold. For each of 5-1, 5-2, and 5-3, the first equation measures θ in radians and the second equation measures θ in degrees.

$$\sin(\theta) = \sin(\theta + 2\pi n), \text{ or } \sin(\theta) = \sin(\theta + 360n) \tag{5-1}$$

$$\cos(\theta) = \cos(\theta + 2\pi n), \text{ or } \cos(\theta) = \cos(\theta + 360n) \tag{5-2}$$

$$\tan(\theta) = \tan(\theta + 2\pi n), \text{ or } \tan(\theta) = \tan(\theta + 360n) \tag{5-3}$$

Example 5-1

Reduce an angle of 1,234° to its reference angle.

Solution. Divide the angle by 360° to express it in terms of rotations, and then convert the fractional rotation back to degrees:

> evalf(1234/360);

```
3.427777778
```

> 360*0.427777778;

```
154.0000001
```

The decimal approximation for the angle is 154 degrees. As a check, the exact answer can be found using *irem*. Here, we find the remainder when 1,234 is divided by 360.

> irem(1234, 360);

```
154
```

Both methods yield 154 degrees.

Your Turn. Reduce 12,345° to its reference angle.

Answer: ______________________________

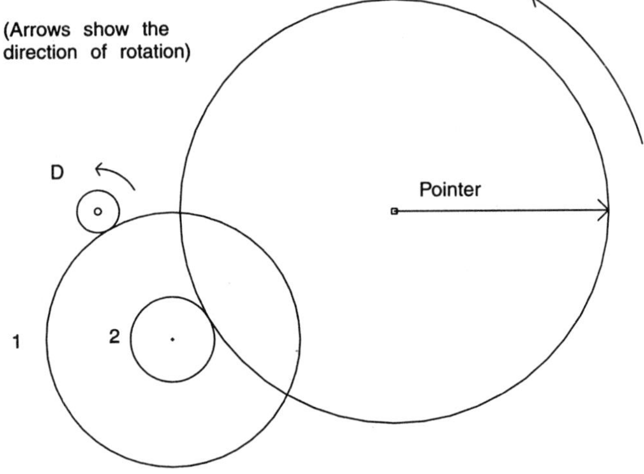

Figure 5.3 A Gear Train

Example 5-2

Large angles are often a part of calculations of the position of a gear in a gear assembly. A large mechanical advantage can be achieved if the driver gear (*D*) is much smaller than the driven gear. In this problem, the driven gear has a pointer attached. The gear ratios are as follows: driver gear—21 teeth, gear 1—155 teeth, gear 2—45 teeth, driven gear—244 teeth. Through what angle does the pointer move if the driver gear makes 10 rotations? Gears 1 and 2 are firmly attached together on the same shaft (see Figure 5.3).

Solution. You build the solution in Maple by alternately counting revolutions and gear teeth. If D makes 10 revolutions, then 10×21 of its teeth are involved. Since it meshes with gear 1, a total of 210 of its teeth are involved. Therefore it rotates by the fraction 210/155. Gear 2 rotates with gear 1, so it makes the same number of rotations: 210/155. Its teeth engage the gear with the pointer. To calculate the number of gear teeth involved, take the number of revolutions of gear 1 (210/155), and multiply by the number of teeth per revolution (namely, 45). This is the number of gear teeth involved for gear 2. This number of gear teeth must also be involved in the pointer gear. The angle must be the ratio of the number of revolutions to the gear teeth in a full revolution, times 360 to convert to degrees. Here is the full calculation (the sequence of numbers alternately counts revolutions, gear teeth, revolutions, gear teeth, revolutions, convert to angle):

> 10*21/155*45/244*360;

$$\frac{170100}{1891}$$

Note that this number is very close to 90º. If the denominator of the fraction were 1,890 instead of 1,891, we would have:

> 170100/1890;

$$90$$

Thus the angle of rotation is almost 90°. Convert 10*21/155*45/244*360 to decimal.

Answer: The precise angle through which the pointer turned is _________ degrees.

Your turn. If the driven gear is changed to one with 122 teeth, and the driven gear makes 20 revolutions, through what angle will the pointer turn?

Answer: ____

Solution of Right-Angled Triangles

Let us review the development we have seen so far.

1. We have defined an angle using the terms *vertex*, *initial side*, and *terminal side.* The initial side lies along the positive x axis.
2. This angle may include any number of full rotations. The remaining angle is called the *reference angle.*
3. A perpendicular line is drawn from the end of the terminal side to the x axis. This construction forms a right-angled triangle.
4. The trigonometric functions for this angle are defined from this right triangle (called the *reference triangle*).

Figure 5.4 shows the reference triangle for an angle of 1,290°. This angle is equivalent to three full rotations plus an additional 210° (1290° = 3 × 360° + 210°). The angle θ'' in the diagram is what other textbooks call the reference angle because of its close relationship to the reference triangle. Examine the right triangle in Figure 5.4. We label the vertical side y, the horizontal side x, and the hypotenuse r. Thus, we can apply the definitions of sine, cosine, tangent, and so forth, to this triangle to find the trigonometric functions for the general angle θ. In our example here, θ = 1,200°. Remember that you must take the algebraic signs of x and y into account when forming the trigonometric ratios. In the example here, both x and y are negative. The hypotenuse, r, is always positive. Maple was helpful enough to provide

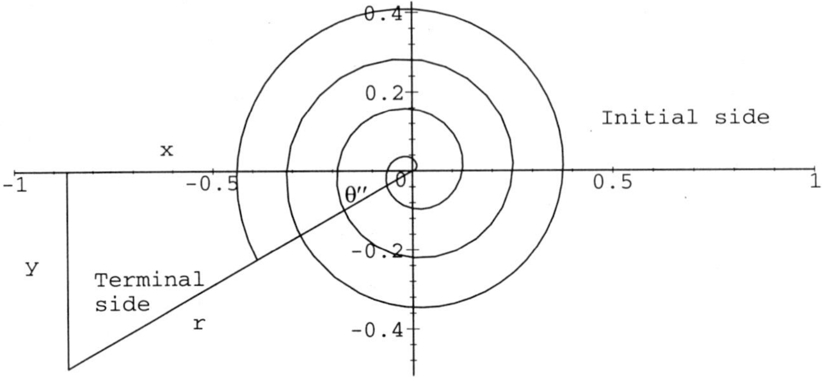

Figure 5.4 The Reference Triangle for an Arbitrary Angle

scales on the x and y axes, so we can estimate the values for the ratios of the sides. The hypotenuse, r, which is also the terminal side, has unit length. The x value is about –0.87, and the y value is about –0.5 (the y tickmarks each count as 0.04). We can write the estimates for sin(θ) and cos(θ) by inspection: sin(θ) = y/r = –0.5/1 = –0.5, and cos(θ) = x/r = –0.87/1 = –0.87. The value for tan(θ) is easily calculated: tan(θ) = y/x = –0.5/–0.87 = 0.57. These estimates are quite close to the actual values. Prove this to yourself by using your calculator and Maple:

sin(1,290°) = ________________________________

cos(1,290°) = ________________________________

tan(1,290°) = ________________________________

The Reference Triangle

You have seen how to construct the reference triangle for any given angle. There is another way of constructing the reference triangle, which is so obvious that it is often overlooked. *The reference triangle can be constructed given the value of any trigonometric function as long as you know which quadrant the angle is in.* Let us say that you know that sin(θ) = 3/5 and θ is in Quadrant 1. Since sin(θ) = y/r, you identify 3 with y and 5 with r, then use the Pythagorean relationship, $x^2 + y^2 = r^2$, to solve for x. Since 16 + 9 = 25, x must be equal to 4. All three sides of the triangle are now known, so the triangle can be constructed. This determines the angle θ. You could use a protractor to measure it. Realize also that any other similar triangle could be used. Usually we choose the hypotenuse to be 1, that is, we set r = 1. This construction is shown in Figure 5.5. All lengths were scaled by dividing each side

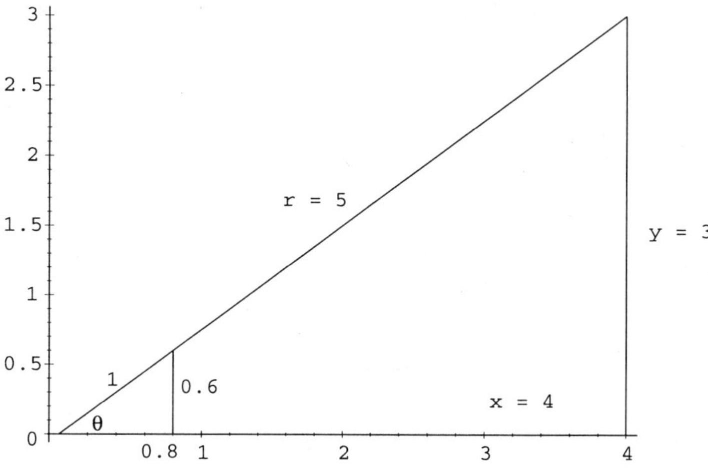

Figure 5.5 The Construction of the Reference Triangle, Given $\sin(\theta) = \frac{3}{5}$

by 5. No matter which similar triangle is used, you can determine every other trigonometric ratio from the diagram. *This is why the method is worth remembering.*

You do not need to have the trigonometric function value expressed as a ratio. Example: Construct the reference triangle for $\sin(\theta) = 0.64$. The simplest solution is to choose the hypotenuse of the reference triangle to be 1. From the definition of the sine function, the altitude of the right triangle will be $y = 0.64$. The base, x is found using the Pythagorean theorem:

$$x = \sqrt{1 - 0.64^2} = \sqrt{0.5904} = 0.768 \tag{5-4}$$

The value of $\cos(\theta)$ is 0.768, and $\tan(\theta)$, which is y/x, is $0.64/0.768 = 0.83$. The other three trigonometric functions are found by taking the reciprocals of these results.

Note that you can scale the numbers for graphical work. For example, if you multiply each value by 10, $y = 6.4$, $x = 7.68$, and $r = 10$. These values are easy to plot on graph paper, and because similar triangles are used, the angle will be the same.

You are now in a position to solve virtually any problem involving the right-angled triangle. The solution of right triangles was covered in Chapter 3, and a few more examples will be given next.

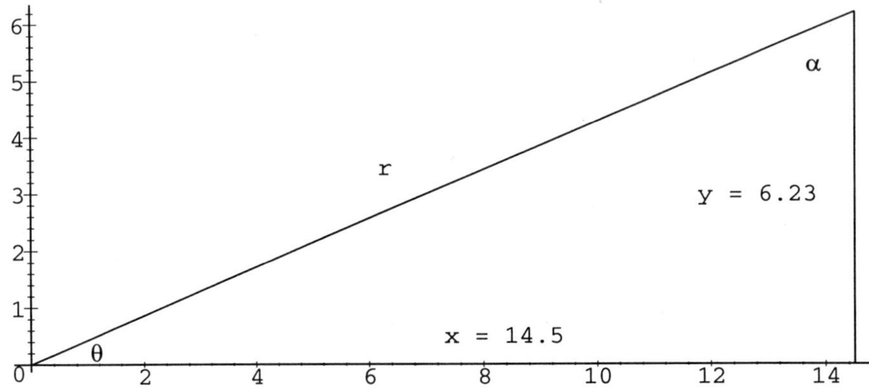

Figure 5.6 Graph for Example 5-3

Example 5-3

Solve for r, θ, and α in the right triangle given in Figure 5.6. Round the answers to three significant figures.

Solution. The full solution to a right-angle triangle problem means finding all the missing parts, in this case, r, θ, and α. Use the Pythagorean relationship to find r.

$$r = \sqrt{x^2 + y^2} = \sqrt{14.5^2 + 6.23^2} = \sqrt{249} = 15.8 \quad (5\text{-}4a)$$

$$\tan(\theta) = \frac{6.23}{14.5} = 0.430,\ \tan(\alpha) = \frac{14.5}{6.23} = 2.33$$

$$\sin(\theta) = \frac{6.23}{15.8} = 0.394,\ \sin(\alpha = \frac{14.5}{15.8} = 0.918$$

$$\cos(\theta) = \frac{14.5}{15.8} = 0.918,\ \cos(\alpha) = \frac{6.23}{15.8} = 0.394$$

The angles θ and α can be found by measuring the diagram with a protractor ($\theta = 23.2^\circ$ and $\alpha = 66.8^\circ$). The angles can be determined using the inverse trigonometric functions, which will be covered in Chapter 9. Remember a theme that goes through this book: if you can't solve a problem by a formula, you can still solve it by applying concepts you do know. Place a protractor on the diagram, and you will measure the angle θ as 23°. You could try estimating the angle if you don't have a protractor handy. Say you estimated the angle as 30°. Now you can find the result, correct to three significant figures, by the trial-and-error method. Calculate $\sin(\theta)$ for $\theta = 30^\circ$, and then revise your estimate of the angle based on the result.

We will use Maple for the trial-and-error calculations. Remember, θ is in degrees, while Maple uses radians! We convert to radians and use *evalf* to convert the answer to decimal:

> evalf(sin(30*Pi/180));

```
.5000000000
```

You see that the result is too high. Now comes the easy part: place Maple's cursor back on the input line and change the number 30 to some smaller number (you might want to try 15). Since you can press Return from anywhere on an input line, do that after revising your estimate. This time, the result will be about 0.26, which is too low, as we are looking for $\sin(\theta) = 0.394$. You will quickly arrive at an estimate of $\theta = 23°$, whose sine is 0.39. A couple more trials should be all that are necessary to achieve three-figure accuracy. Your result will be $\theta = 23.2°$, making $\alpha = 66.8°$. The point we are making here is that it is *easy* to apply the knowledge you already have to solve this problem without knowing the use of the inverse trigonometric functions. Take the straightforward approach of using the definition of the sine function: type it in on a Maple input line. Since you don't know the angle, *guess*! When you execute the command, you will see whether you are right. Since it is so easy to edit the command, you can perform new trials quickly and effortlessly. It's also good preparation for understanding the inverse functions, so that when you encounter them in Chapter 9, you'll be ready.

Your Turn.

(a) Solve the right triangle if $x = 72$ and $y = 35$. *Answer:* ____________

(b) Solve the right triangle if $r = 127$ and $\theta = 155°$. *Answer:* ____________

(c) Solve the right triangle if $r = 0.154$ and $\theta = 295°$. *Answer:* ____________

(d) Solve the right triangle if $x = -42.5$ and $y = -62.7$. *Answer:* ____________

Example 5-4

A surveyor makes two sightings of a mountain using his transit. His station is above sea level, but the mountain starts at the sea and rises inland. He records his readings in Figure 5.7. Since both angle measurements were made with respect to a horizontal line, two right-angled triangles are formed.

The baseline for this problem is 1.5 mi, or 7,920 ft. Thus, side *a* is

> a = evalf(7920*tan(Pi/180*(6+16/60)))*ft; (Note how 16′ is converted to degrees.)

```
a = 869.71261 ft
```

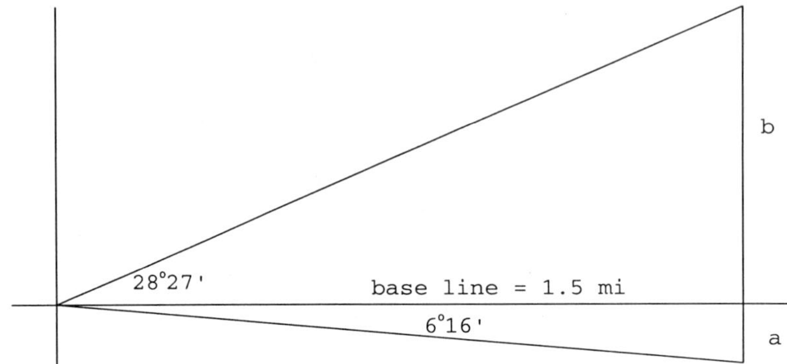

Figure 5.7 A Surveying Problem

```
> b = evalf(7920*tan( Pi/180*(28+27/60)))*ft;
```

$$b = 4291.264356 \ ft$$

```
> a + b = (869.7 + 4291)*ft;
```

$$a + b = 5161 \ ft$$

Look at the Maple commands that were used to solve the problem and note these points:

1. Maple uses radian measure, so angles in degrees are multiplied by $\pi/180$.
2. Surveying angles are measured in degrees and minutes. Minutes are converted to fractional degrees by dividing by 60.
3. The *evalf* command is used to convert answers to decimal.
4. Maple statements (as shown in this problem) containing an equal sign are *not* assignments. Thus, the values of *a* and *b* in the first two commands are not remembered by Maple.
5. Expressions were multiplied by the unevaluated name *ft* so that the answers would contain the physical unit as well as the numerical value. This is simply a matter of style.

The main point to keep in mind is the way in which you build up a solution using a Maple worksheet. The next example shows more of this approach.

Your Turn.

(a) What are the distances a and b if the angle above the horizontal is $33°50'$ and the angle below the horizontal is $10°17'$?

Answer: ______________________________

(b) If $a = 560$ ft and $b = 4{,}250$ ft, what are the angles in degrees and minutes?

Answer: ______________________________

Example 5-5

A triangle has sides of 6.14 m, 7.28 m, and 12.4 m, respectively. Find $\cos(\theta)$, where angle θ is shown in Figure 5.8.

Solution. The problem, as originally stated, does not refer to a right-angled triangle. Constructing the perpendicular line AB forms two right-angled triangles, ABC and ABO. This construction is key to finding a solution to this problem. Sometimes you have to "complicate" a problem to solve it! Drawing construction lines is a powerful technique in geometrical problem solving (see Figure 5.8).

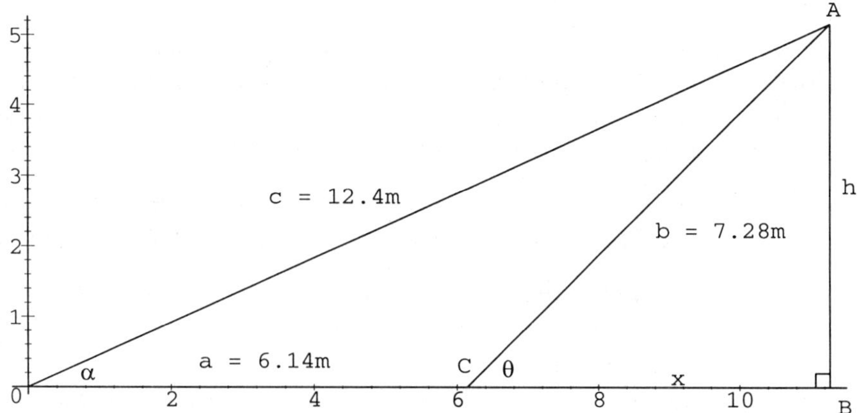

Figure 5.8 Constructing Right-Angled Triangles as a Method of Solution

Applying the Pythagorean theorem to triangle ABO, we have:

> eq1 := c^2 = h^2 + (a + b*cos(theta))^2;

$$eq1 := c^2 = h^2 + (a + b\cos(\theta))^2$$

We have used the relation $x = b\cos(\theta)$, since the base of triangle ABO is $a + x$. Next, we *expand* the term containing $\cos(\theta)$ and we *subs*titute for h at the same time.

> eq2 := expand(subs(h=b*sin(theta), eq1));

$$eq2 := c^2 = b^2\sin(\theta)^2 + a^2 + 2b\cos(\theta)\ a + b^2\cos(\theta)^2$$

This expression can be simplified:

> eq3 := simplify(eq2);

$$eq3 := c^2 = a^2 + b^2 + 2b\cos(\theta)\,a$$

Finally, we solve for $\cos(\theta)$ and substitute the given values of a, b, and c:

> with(student): eq4 := isolate(eq3, cos(theta));

$$eq4 := \cos(\theta) = -\frac{1}{2}\,\frac{-c^2 + a^2 + b^2}{ba}$$

> subs(a= 6.14, b=7.28, c=12.4, eq4);

$$\cos(\theta) = .705404123$$

We could use the method of trial and error, described in Example 5-2, to find that $\theta = 45.14°$.

Look at the Maple commands that were used to solve the problem and note these points:

1. Even though the problem as stated did not contain a right angle, a construction line was used to create two right-angled triangles.
2. We wrote down an equation, based on the Pythagorean theorem, which contained $\cos(\theta)$. Unfortunately, it contained $\sin^2(\theta)$ and $\cos^2(\theta)$ as well. Since we have used the Maple command *simplify* before, we tried it to see what would happen. The $\sin^2(\theta)$ and $\cos^2(\theta)$ disappeared! We will see why in Chapter 6.
3. We used the student package from Maple to use the *isolate* command. We could have used *solve* as well. In many cases, *solve* and *isolate* do essentially the same job.
4. We found the value of $\cos(\theta)$ by substitution. By organizing the work in this way we have a record of all the steps leading to the solution of our problem. We could save the result to disk so that any time we wanted to solve a problem giving the three sides of a triangle, we would have a model to use. You have probably noticed that we have managed to *derive* the famous *cosine law* formula as a "side effect" of solving this problem. In other words, we have found a formula that can be used any time we are

given the three sides of a triangle. The formula allows us to calculate the exterior angle at the base. Knowing this, the interior angle is found by subtracting the exterior angle from 180°. Note that the angle α could have been found by trial and error also. The formulas will be derived again in Chapter 9.

Your Turn.

(a) Find $\cos(\theta)$ if $a = 70$, $b = 80$, $c = 130$. *Answer:* ____________

(b) Find $\sin(\theta)$ if $a = 0.527$, $b = 0.650$, $c = 1.03$. *Answer:* ____________

Periodic Functions

Figures 5.1 and 5.4 show that the trigonometric functions have the same value when the angle's terminal side is given any number of additional full rotations. Since the terminal side lies over itself once more, after any number of additional full rotations, the trigonometric ratios calculated from it must be the same. Each rotation is called a *cycle.* The trigonometric ratios repeat themselves every cycle. When a function behaves in this manner, it is called a *periodic function.* This relationship of the trigonometric function to itself can be expressed by the following formulas (in radians):

$$\sin(\theta + 2\pi n) = \sin(\theta),\ n = 0,1,2,3\ldots \tag{5-5}$$

$$\cos(\theta + 2\pi n) = \cos(\theta),\ n = 0,1,2,3\ldots \tag{5-6}$$

$$\tan(\theta + 2\pi n) = \tan(\theta),\ n = 0,1,2,3\ldots \tag{5-7}$$

This last relation requires some clarification. You know that if you rotate the terminal side of the angle through an additional full rotation, it will lie over the top of its original position. The tangent will have the same value before and after the full rotation. Could this happen sooner than one full rotation? In the case of the tangent function, it can. Examine Figure 3.6 carefully. It shows that the tangent function repeats itself every time the terminal side is increased by π radians, or 180°. Since both x and y are negative in quadrant 3, the tangent function will be the same as it was in quadrant 1. In other words, $\tan(\pi + \theta) = \tan(\theta)$, so the tangent function repeats itself twice as often as the sine and cosine functions. So you have the even stronger relation

$$\tan(\theta + \pi n) = \tan(\theta),\ n = 0,1,2,3\ldots \tag{5-8}$$

Of course, this relation is perfectly consistent with Eq. 5-7. Eq. 5-8 is the stronger relation.

Maple automatically applies this simplification if the number of rotations is given explicitly in an expression. Observe the following results of evaluating some trigonometric expressions. Assume that α, β, and γ are all less than a full rotation.

```
> sin(2*Pi + alpha), cos(24*Pi + beta), tan(3*Pi + gamma);
```

$$\sin(\alpha),\ \cos(\beta),\ \tan(\gamma)$$

In each case, Maple knew to reduce the angle to the reference angle. It discarded the extra rotations, or, in the case of the tangent function, the extra half-rotations. Note, however, what happens if we ask Maple to evaluate the sine function when we give the reference angle an extra half-rotation:

```
> sin(3*Pi + alpha);
```

$$-\sin(\alpha)$$

The result is the negative of the sine of alpha! Can you see why? If you add an extra half-rotation, the terminal side will be opposite where it was. This will cause the sign change. (The same is true for the tangent function, but since both x and y change sign, the result will remain the same. With the sine or cosine function, r is always positive so there is only one sign change, resulting in the negation of the original value.)

Definition of Amplitude, Period, Frequency, and Phase

The periodicity of the sine and cosine functions are useful, especially because of their close relationship to circular motion (Figure 3.2). Many oscillating physical systems can be modeled by a sine or cosine function whose angle varies linearly with time. Such functions are sometimes called *sinusoids*. The motion of a child on a swing, a pendulum clock (grandfather clock), a vibrating drum, and a diatomic molecule can all be described by sinusoids. In electricity, sinusoids are central to the theory of alternating-current circuit theory.

These sinusoids are usually functions of time, so we will use a sine function whose angle varies with time in the following discussion. The type of motion we want to discuss has the property that the excursions repeat over and over: that is, the motion is periodic. Diatomic molecules may vibrate billions of times a second, or even faster. The extent of their motion will be very tiny, however. A child on a swing will sway back and forth every second or so, and the extent of the motion will be a few feet. An electric signal may oscillate very slowly (as in undersea submarine communications) or incredibly quickly (as in a microwave oven or satellite communications). In all these cases, the general expression looks like:

$$f(t) = A\sin(2\pi ft + \alpha) \tag{5-9}$$

The quantity A is called the *amplitude*, f is the frequency, and α is the *phase*. The variable t measures the time in some convenient unit. If we replace $2\pi f$ by the single variable ω, the formula becomes:

$$f(t) = A\sin(\omega t + \alpha) \tag{5-10}$$

We know that this function is periodic so we ask when it will repeat itself. The answer, of course, is: when the angle increases by 2π radians. (Radian measure is used almost exclusively in studying dynamic systems.) The value of the function at $t = 0$ is $f(0) = A\sin(\alpha)$. At some later time T, one full oscillation of the sine function will have occurred.

$$f(T) = f(0) = A\sin(\omega T + \alpha) = A\sin(\alpha) \tag{5-11}$$

The only way to get the function to execute a full oscillation is to have the angle increase by 2π. Thus we have the equation:

$$\omega T = 2\pi \tag{5-12}$$

$$T = \frac{2\pi}{\omega} \tag{5-13}$$

The time, T, is called the *period* of the oscillation. It is the time it takes to complete one full oscillation. The reciprocal of this time is called the *frequency*. The frequency measures the number of complete oscillations per unit of time. This implies the relation:

$$\frac{1}{f} = \frac{2\pi}{\omega} \tag{5-14}$$

$$\omega = 2\pi f \tag{5-15}$$

The quantity ω is called the *angular frequency*. The quantity α is called the *phase*. You could arrange things so that you start your timing device when the sinusoid crosses 0 on an upward swing. It is often easier to measure the angle of the sinusoid when $t = 0$ and call this angle the *phase shift*.

A picture really helps here (see Figure 5.9). We want to compare the two sinusoids:

$$f_1(t) = A\sin(\omega t), \text{ and } f_2(t) = A\sin(\omega t + \alpha) \tag{5-16}$$

for some specific values of A, ω, and α. In $f_1(t)$, let $A = 10$, $\omega = 1$, and $\alpha = 0$. In $f_2(t)$, let $A = 10$, $\omega = 1$, and $\alpha = 0.03$ radian. The graph of $f_1(t)$, with t ranging from 0 to 6.28 s, is

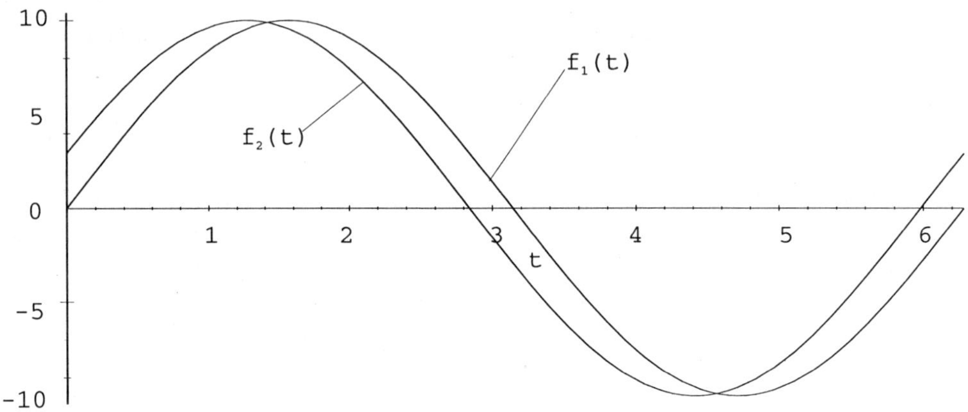

Figure 5.9 Two Sinusoids: Curve $f_2(t)$ Is Phase Shifted Compared to Curve $f_1(t)$

> f1 := 10*sin(t); f2 := 10*sin(t + 0.3);

```
f1 := 10 sin(t)
f2 := 10 sin(t + .03)
```

> plot({ f1, f2 }, t = 0 .. 6.28);

Look at the Maple commands used in this example and note these points:

1. The angular frequency, ω, is 1 radian per second. Consequently, a full wave will take 2π seconds. This is why we used 6.28 s in the range for time ($2\pi \approx 6.28$).

2. The phase shifted curve, $f_2(t)$, crosses the y axis *before* the graph of $f_1(t)$. *A positive phase angle shifts the plot to the left.* There is no other change to the curve. The amplitude and the frequency remain the same. It is as if you were able to pick up the curve and slide it in the direction of the negative x axis.

3. The amount moved is equivalent to a *phase shift*. This is different from the phase. The way we are using the term here, the phase shift causes the sine wave to move over a certain "distance" (really, a time interval) along the t axis, whereas the phase is an angle. The phase shift is given by α/ω. When the angular frequency is unity, as it is in this case, the phase and the phase shift are numerically the same. If $\omega \neq 1$, the phase shift must be calculated. In the example here, the phase shift is 0.3 seconds, whereas the phase is 0.3 radians.

4. We defined the two expressions by giving them the *names* f1 and f2. This way, we could use them in the plot command. Maple replaces these names with their values and plots the result. The values are the two sinusoids. Note the Maple syntax using curly braces for plotting a *set* of functions (two in this case).

Examples and Solved Problems

Example 5-6

Two sinusoids both have amplitudes of 10 units. One has an angular frequency of 2 s^{-1} and the other has an angular frequency of 10 s^{-1}. Plot both on the same graph for one full oscillation of the slower wave. Both sinusoids have a phase of 0.

Solution. You do not need to define the two functions; you can write the formulas directly into the *plot* command. The slower wave has an angular frequency of 2, so the range should be $t = 0$ .. Pi.

> plot({ 10*sin(2*t), 10*sin(10*t) }, t = 0 .. Pi);

See Figure 5.10. Note that the fast wave, whose angular frequency is five times that of the slow wave, has five peaks in the same region where the slow wave has one.

Your Turn. Sinusoid *A* is given by:

> A := 3*sin(4*t + Pi/10);

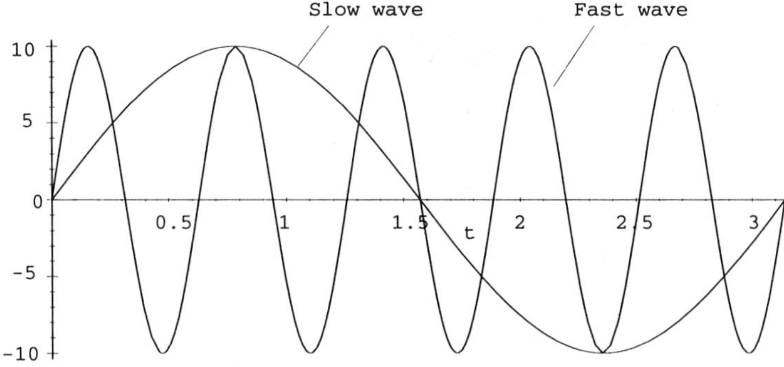

Figure 5.10 Two Sinusoids of the Same Amplitude But Different Frequencies

and sinusoid *B* by:

> B := 4*sin(3*t-Pi/10);

(a) Plot both sinusoids on the same graph for $t = 0$ .. Pi. Find the phase shift (in seconds) between the two waves (the shortest time between *A* crossing the *t* axis and *B* crossing the *t* axis).

Answer: ____________________

(b) Does either sinusoid start at the origin (0, 0)? *Answer:* ____________

Example 5-7

Add the waves $f_1(t) = 3*\sin(t)$ and $f_2(t) = 4*\sin(t + 0.25)$. Choose a plot interval so that one full wave from $f_1(t)$ and one from $f_2(t)$ are shown. What can you conclude about the shape of the resulting waveform?

Solution. Plot both waves and their sum on the same graph (see Figure 5.11). Add the phase shift of $f_2(t)$ to the beginning of the interval so that a complete oscillation of both waves will be shown.

> plot({ 3*sin(t), 4*sin(t + 0.25), 3*sin(t) + 4*sin(t + 0.25) }, t = -0.25 .. 2*Pi);

The sum of the two sine waves looks like a sine wave, too. It is not obvious how to perform this sum algebraically, but the experimental evidence from the graph is quite convincing. You will see a derivation in Chapter 8.

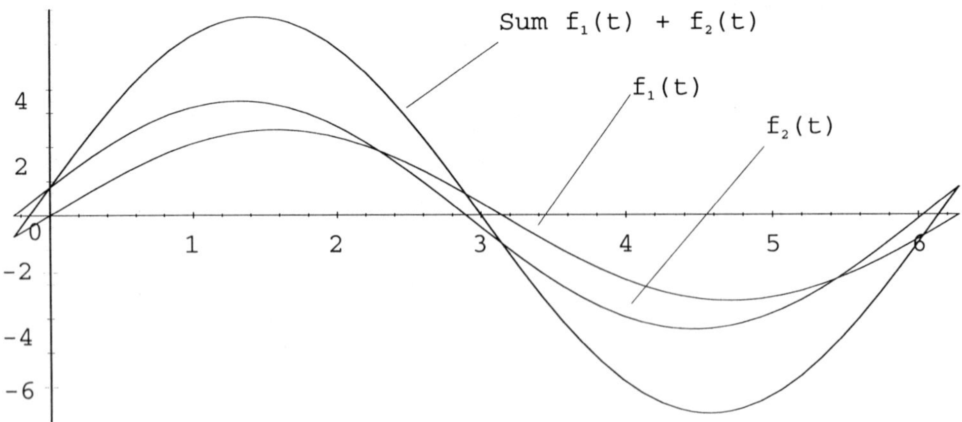

Figure 5.11 The Sum of Two Sinusoids

Your Turn. Add sinusoids A and B where:

> A := 5*sin(10*t);

and:

> B := 4*sin(10*t + Pi);

Display A, B, and $A + B$ on the same plot, over the interval $t = 0$ .. Pi/5.

(a) Compare the sum to the sinusiods making up the sum:

Answer: ______________________________

(b) Is $A + B$ greater or less than A; than B? *Answer:* ______________

Example 5-8: Cofunction Relationship of Sin(θ) and Cos(θ)

Plot sin(θ), cos(θ) and sin(θ+π/2) on the same graph. What can you conclude about the relationship between sin(θ+π/2) and cos(θ)?

Solution. (See also Figure 5.12):

> plot({ sin(theta),cos(theta),sin(theta+Pi/2) }, theta = 0 .. 2*Pi);

The two curves, cos(θ) and sin(θ+π/2), have the same plot. The only way to tell that sin(θ+π/2) is really there is to plot it by itself. Do this yourself to prove that it really has the same plot as cos(θ). You will be asked to show other relationships of this type in the Maple exercises for this section.

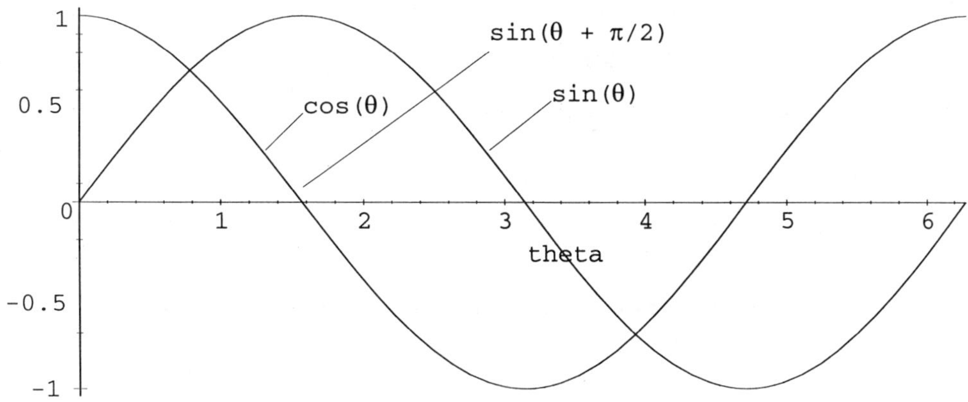

Figure 5.12 Plot of sin(θ),cos(θ) and sin(θ+π/2)

Paper and Pencil Exercises

PP5–1

Convert the following revolutions to degrees.

(a) 21 1/2 rotations = _________ degrees

(b) 66 2/3 rotations = _________ degrees

(c) 2π rotations = _________ degrees

PP5–2

Convert the following angles to rotations. State your answer in full rotations and the remainder in degrees.

(a) 3,600 degrees = _________ rotations

(b) 1,450 degrees = _________ rotations

(c) 1,890 degrees = _________ rotations

(d) 2,700 degrees = _________ rotations

(e) 4,000 degrees = _________ rotations

PP5–3

Convert the following rotations to radians. Give the *exact* answer.

(a) 60 rotations = _________ radians

(b) 2π rotations = _________ radians

(c) $\pi/2$ rotations = _________ radians

PP5–4

Convert the following radian measures to rotations. Give your answers in decimal form.

(a) 20π radians = _________ rotations

(b) 23 radians = _________ rotations

(c) 99 radians = _________ rotations

PP5–5

How many full rotations is:

(a) $720^\circ =$ _________ rotations

(b) 36,000 radians = _________ rotations

(c) $1080^\circ =$ _________ rotations

(a) 40π radians = _________ rotations

(b) 7 radians = _________ rotations

(c) 62,800 radians = _________ rotations

PP5–6

By subtracting full rotations, state the angle less than 360° (or 2π radians) to which these angles are equivalent. In other words, state the reference angle:

(a) $420^\circ =$ _________ degrees

(b) 3π radians = _________ degrees

(c) $1{,}333^\circ =$ _________ degrees

(d) $855^\circ =$ _________ degrees

(e) 9.87 radians = _________ degrees

(f) 1,333 radians = _________ degrees

PP5–7

You might want to use Maple to check this one: how many degrees does the earth turn through in a year? How many radians does it turn through? First, state the problem more precisely. There are at least two possible interpretations:

Interpretation of the problem: ____

____

____

Answer: The earth turns through _________ degrees, or _________ radians.

PP5–8

On paper, construct the reference triangle for $\cos(\theta) = 0.1$. Measure the angle with a protractor:

Answers: Angle in degrees____________ Angle in radians____________

Maple Lab

ML5–1: Cofunction Relationships

Demonstrate these cofunction relationships by plotting the functions in Maple.

(a) $\sin(x) = \cos(90°-x)$ (b) $\cos(x) = \sin(90° - x)$ (c) $\tan(x) = \cot(90° - x)$

(d) $\cot(x) = \tan(90° - x)$ (e) $\sec(x) = \csc(90° - x)$ (f) $\csc(x) = \sec(90° - x)$

By examining the graphs of the two functions, determine the relationships between:

(a) $\sin(x)$ and $\cos(90° + x)$ *Answer:* ____________________

(b) $\cos(x)$ and $\sin(90° + x)$ *Answer:* ____________________

(c) $\sin(x)$ and $\sin(180° + x)$ *Answer:* ____________________

Remember to convert degrees to radians before plotting the functions!

ML5–2

Adding two sinusoids of different frequency gives a much different result from Example 5-7, where both sinusoids had the same frequency. Plot the expression $3\sin(2x) + 2\cos(3x)$, along with $3\sin(2x)$ and $2\cos(3x)$. Is the graph of $3\sin(2x) + 2\cos(3x)$ a sinusoid? Describe the shape of the resultant wave:

Answer: ____

ML5–3: Functional Values

You will need to evaluate "functions of functions" in trigonometry, and you will need to do so again in calculus. These calculations are simplified by using functional notation.

Define the functions

```
> f := theta -> 1 + sin( 2*theta );
```

$$f := \theta \rightarrow 1 + \sin(2\theta)$$

> g := theta -> 2 * cot(Pi/4 * (theta-1));

$$g := \theta \rightarrow 2\cot\left(\frac{1}{4}\pi(\theta - 1)\right)$$

(a) We have used the name θ in both definitions. Is there a naming conflict here? (The answer is no.) In Maple, issue the following commands and answer the questions.

(i) Evaluate > f(2*Pi); *Answer:* ____________

(ii) Evaluate > g(2); *Answer:* ____________

(iii) Evaluate > f(x); *Answer:* ____________

The advantage of this notation is the ease with which you can evaluate the functions at various points.

(b) Evaluate:

(i) > f(Pi/3); *Answer:* ____________

(ii) > g(3); *Answer:* ____________

Check Maple's answers by evaluating these functions with your pocket calculator. Once you have typed in the commands for $f(\theta)$ and $g(\theta)$, it is easy to replace θ with any value you choose. You can obtain the new results with a simple editing operation.

(c) Evaluate $g(1)$. Explain Maple's result:

Answer: ____________

(d) Here you will evaluate a "function of a function." Type in the command

> g (f (Pi/12));

and explain the result. First, the function, f, is evaluated at $\theta = \pi/12$. Call the result x. The function $g(x)$ is evaluated to obtain the final result.

Answer: ____________

(e) Investigate these commands:

(i) f(Pi/4); *Answer:* ____________

(ii) g(f(Pi/4)); *Answer:* ____________

(iii) g(g(f(Pi/4))); *Answer:* ____________

(iv) g(g(g(f(Pi/4)))); *Answer:* ____________

Explain these results:

Answer: ____________

(f) Evaluate, and interpret your results:

> f (1.377336877);

Answer: f = ______________________________

Explanation: ______________________________

(g) Try other combinations of *f* and *g*. Write down some of your more interesting results. For example, compare the result you get from *f*(*g*(3)) and *g*(*f*(3)).

Answer: ______________________________

Other examples: ______________________________

____

____

ML5–4: Solution of a Right Triangle

Find the angle and the lengths AD and AC (see Figure 5.13).

> ; (Write the Maple commands that solve this problem here) ______________________________

____

Answers:

CB = ______________________________

θ = ______________________________

AD = ______________________________

AC = ______________________________

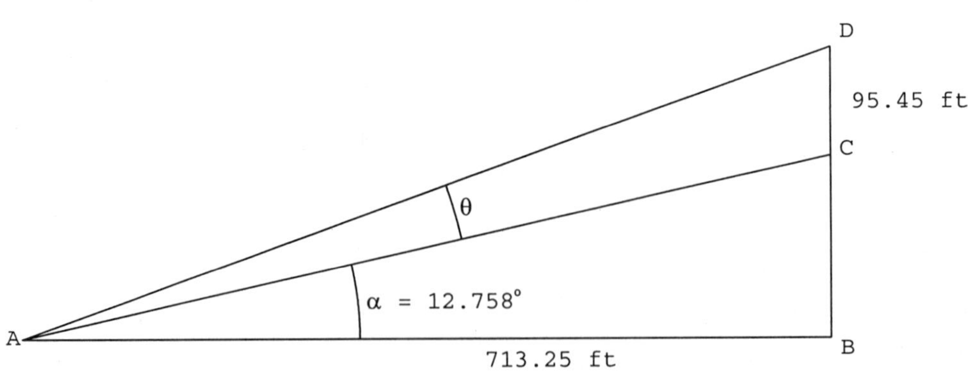

Figure 5.13 Diagram for ML5-4

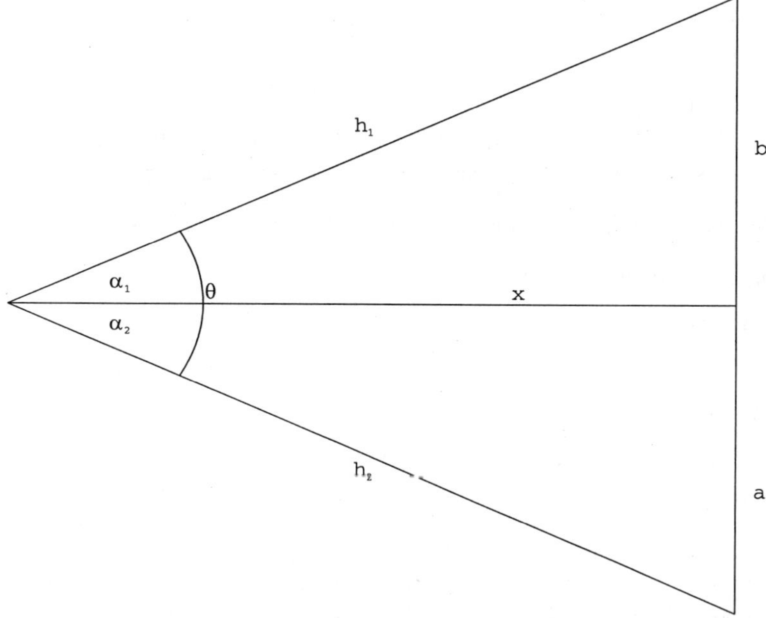

Figure 5.14 Diagram for ML5–5

ML5–5: Solution of a Right Triangle

Given $x = 10.11$ m, $a = 3.47$ m, and $b = 4.23$ m, find θ, h_1, and h_2 (see Figure 5.14).

> ; (Write the Maple commands that solve this problem here)____________________

____

Answers: $y =$ ____

$\alpha_1 =$ ____

$\alpha_2 =$ ____

$\theta =$ ____

$h_1 =$ ____

$h_2 =$ ____

Explorations

E5-1

In Example 5-7 you found the sum of two sine waves. You investigated this type of sum again in ML5–2, where the two sine waves had different frequencies.

Explore what happens when two sine waves are multiplied together. Describe the results of:

(a) Multiplying a sine wave by itself:

> **a := sin(x)*sin(x);**

(b) Multiplying a sine wave by a cosine wave:

> **b := sin(x)*cos(x);**

(c) Multiplying a sine wave by another at higher frequency:

> **c := sin(x)*sin(10*x);**

You might want to begin your explorations by defining the functions you are going to use and then plotting them. See if you can predict the shape of a curve before you plot it. Change the angle variable from θ to 2θ and then to 3θ. See if this makes any essential difference in the plot, other than taking less time to complete a full oscillation.

Record your investigations and your plots. You will return to forms (a) and (b) later on in this text. Form (c) is called *amplitude modulation*.

CHAPTER 6

Basic Trigonometric Identities

Objectives for This Chapter

1. Understand that reciprocal functions are identities
2. Define the fundamental identities and apply them to reduce more complicated identities
3. Use Maple methods for proving identities

Maple Commands Used in This Chapter

solve(eqn, var)	Solve a trigonometric equation *eqn* for variable *var*.
simplify(expr)	Simplify a trigonometric expression.
subs(eqn, expr)	Substitute for a variable in *expr*.

The Difference between an Equation and an Identity

Simplicity of expression is considered a virtue in mathematics. For this reason, mathematicians try to reduce their results to the most economical form. This is as true for trigonometric expressions as it is for algebraic ones. A mathematician would write $3x$ instead of $x + x + x$, and so would you. If you analyzed a diagram containing triangles and you managed to derive the expression, $\frac{\sec(y)^2}{\cot(y)} - \tan(y)^3$, you might not realize immediately that the expression can be simplified. If you were going to use this result in a published report or as the basis for further repetitive work, you would want to use the simplest possible form. This chapter will demonstrate some techniques you can employ to examine expressions of this type to look for possible simplifications.

A related problem is to determine whether two expressions are equivalent. Consider the three expressions:

$$\sec(x)^2 + \csc(x)^2 = \sec(x)^2 \csc(x)^2 \qquad (6\text{-}1)$$

$$\sin(x)^2 - \cos(x)^2 + 1 = 0 \qquad (6\text{-}2)$$

$$\tan(x) = -\cot(x) \qquad (6\text{-}3)$$

Examine these mathematical statements. Each statement says that the right-hand side equals the left-hand side. There are three possibilities. First, the statement may be false, which happens when *no* value of x will make the statement's right-hand side equal to the left-hand side. Here, we have a *contradiction*. Second, the statement may be true for specific values of x and false for other values. Then we have an *equation*, which can be solved for the set of x values that make the equation true. Last, the statement may be true no matter what value of x is used. This type of statement is called an *identity*.

Look again at the three statements. You might be tempted to guess that Equation 6-1 is an equation because it resembles the algebraic equation $x^2 + y^2 = x^2y^2$. The equality is satisfied for some values of x and y, but not for others. If $x = 2$, then $4 + y^2 = 4y^2$, so $3y^2 = 4$, $y = 2/\sqrt{3}$, and we have a solution. You can easily prove to yourself that if $x = 1$, there is no solution. Thus, $x^2 + y^2 = x^2y^2$ is an equation. This is the line of reasoning that might lead you to believe that Equation 6-1 is an equation, but is it?

We will review the fundamental trigonometric identity $\sin(x)^2 = 1 - \cos(x)^2$ in this chapter. This equation is also an identity. Therefore, a quick glance at Equation 6-2 may lead you to believe that it is an identity as well. Additionally, doesn't it make sense that there must be some value of x that makes Equation 6-3 true? Perhaps $x = 0$?

Have you been misled by these arguments into believing that Equations 6-1 and 6-3 are equations, while Equation 6-2 is an identity? In fact, Equation 6-1 is an identity, Equation 6-2 is an equation, and Equation 6-3 is a contradiction. These examples have been chosen to encourage you to accept the wrong conclusion. Our purpose here is to emphasize the need for caution: a casual scan of a formula can be dangerous! It is no substitute for careful analysis. We will show you how analyze Equations 6-1, 6-2, and 6-3 in this chapter. You should take from this example the fact that it is by no means easy to determine whether a trigonometric expression is an equation, an identity, or a contradiction.

Definitions of the Fundamental Trigonometric Identities

From the relationships among the six basic trigonometric functions, the following identities emerge:

$$\sin(\theta) = \frac{1}{\csc(\theta)} \tag{6-4}$$

$$\cos(\theta) = \frac{1}{\sec(\theta)} \tag{6-5}$$

$$\tan(\theta) = \frac{1}{\cot(\theta)} \tag{6-6}$$

$$\tan(\theta) = \frac{\sin(\theta)}{\cos(\theta)} \tag{6-7}$$

All these identities are really definitions; however, the next relationship is derived from an analysis of the right-angled triangle. Since $\sin(\theta) = y/r$ and $\cos(\theta) = x/r$, we have, by squaring both equations, $\sin(\theta)^2 = y^2/r^2$ and $\cos(\theta)^2 = x^2/r^2$. Adding the squared terms yields, $\sin(\theta)^2 + \cos(\theta)^2 = x^2/r^2 + y^2/r^2 = (x^2 + y^2)/r^2$. Since the Pythagorean theorem states that $x^2 + y^2 = r^2$, we have $\sin(\theta)^2 + \cos(\theta)^2 = r^2/r^2 = 1$. Therefore, we have proved one of the most useful identities in trigonometry (see Figure 6.1). We will state it again:

$$\sin(\theta)^2 + \cos(\theta)^2 = 1 \tag{6-8}$$

$$\sin^2(\theta) + \cos^2(\theta) = 1 \tag{6-9}$$

The first form, Equation 6-8, is how it is written in Maple; the second form is completely equivalent. The second form is the form found in mathematics textbooks and in the literature.[1] This identity is of the utmost importance. It must be *memorized* in all its equivalent forms, $\sin(\theta)^2 + \cos(\theta)^2 = 1$, $\sin(\theta)^2 = 1 - \cos(\theta)^2$, $\cos(\theta)^2 = 1 - \sin(\theta)^2$, $\sin(\theta) = \pm\sqrt{1 - \cos(\theta)^2}$, $\cos(\theta) = \pm\sqrt{(1 - \sin(\theta)^2)}$. Read all these equations and make sure that you can mentally transform the first of these into the rest.

Similarly, you can prove that:

$$\tan(\theta)^2 + 1 = \sec(\theta)^2 \tag{6-10}$$

---

1. In Maple, you could write:

> (sin^2)(x);

but Maple would still respond with:

```
sin(x)^2
```

For this reason it is preferable to write, > *sin(x)^2;* There is less chance of confusion. Note that > *sin^2(x);* does not work!

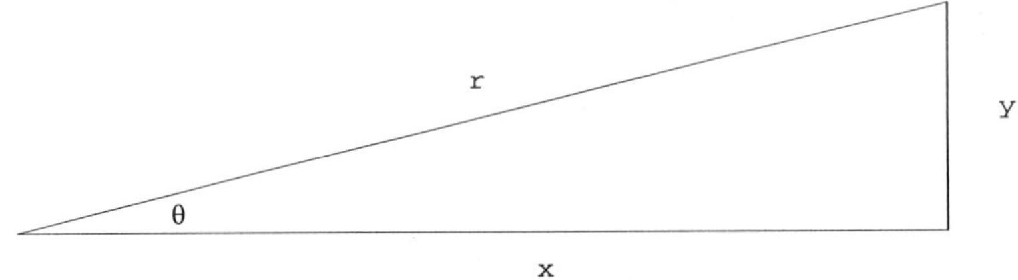

Figure 6.1 Analysis of the Right-angled Triangle

$$\cot(\theta)^2 + 1 = \csc(\theta)^2 \qquad (6\text{-}11)$$

by defining the trigonometric relationships in terms of the sides of the defining triangle (Figure 6.1). The proof will be left to the paper and pencil exercises. These identities must be memorized, too. Trigonometry will always be a struggle without a working knowledge of all the basic relationships discussed so far in this chapter.

The trigonometric functions are so closely related that it is easy to write them in equivalent forms. In other words, trigonometric identities abound. Our purpose in this workbook is to show you techniques for determining if, for any given angle, two trigonometric expressions really are the same. If they are, equating the expressions will result in an identity. You can verify simple identities by paper and pencil methods, as we did to prove that $\sin(\theta)^2 + \cos(\theta)^2 = 1$. However, when the expressions are complex, it is wise to apply computer algegbra methods. Assume that you have one of these rather complex situations and want to see if a trigonometric equation is an identity.

Step 1: Plot both expressions in Maple. If both curves lie over one another, it is strong evidence that the two expressions are identical. It is not a *proof,* because all you can say for sure is that the two expressions are equal to about three significant figures. Consider the two expressions, x^2 and $x^2 + 0.000001$. Equating these expressions is a contradiction, not an identity, yet the two plots are indistinguishable over the range $x = 0 .. 1$.

Step 2. Subtract the two expressions in Maple and simplify the result. Maple is able to reduce many identities to 0 in this way, which works better than simplifying the two expressions separately. They usually yield different results in Maple. The results are equivalent, but they do not appear to be equivalent. Since Maple works symbolically, a result of 0 shows that the expressions are truly identities.

These methods will be demonstrated in the next section.

Example 6-1

Prove that $\dfrac{\tan(y)}{\sin(y)} = \sec(y)$

Solution. Type the equation in Maple:

> e1 := tan(y)/sin(y) = sec(y);

$$e1 := \frac{\tan(y)}{\sin(y)} = \sec(y)$$

> simplify(e1);

$$\frac{1}{\cos(y)} = \frac{1}{\cos(y)}$$

In this case, the *simplify* command works without the need to do any further manipulations. Maple prefers to give its result in terms of the basic trigonometric functions, not their reciprocals, so it does not output sec(y) = sec(y). Nonetheless, Equation $e1$ is shown to be an identity.

Your Turn. Prove that $\dfrac{\cot(y)}{\csc(y)} = \cos(y)$ is an identity.

Answer: ______________________________

Example 6-2

Prove that $\sec(\theta)\tan(\theta)\csc(\theta) = \tan(\theta)^2 + 1$.

Solution. Type the equation in Maple:

> e2 := sec(theta)*tan(theta)* csc(theta) = tan(theta)^2 + 1;

$$e2 := \sec(\theta)\tan(\theta)\csc(\theta) = \tan(\theta)^2 + 1$$

Note that by giving the original equation a name, you can check Maple's output against the statement of the problem. Now you can work with the equation with the knowledge that it is correct.

Try simplifying the equation:

> simplify(e2);

$$\frac{1}{\cos(\theta)^2} = \frac{1}{\cos(\theta)^2}$$

It still works! Maple shows you that the equation is indeed an identity. Some people may think this is "cheating," because Maple did all the work. If you feel this way, by all means do the problem by paper and pencil methods as well. Remember that $\tan(\theta)^2 + 1 = \sec(\theta)^2$? The equation can be rewritten:

$$\sec(\theta)\frac{\sin(\theta)}{\cos(\theta)}\frac{1}{\sin(\theta)} = \sec(\theta)^2 \tag{6-12a}$$

$$\frac{\sec(\theta)}{\cos(\theta)} = \sec(\theta)\,\sec(\theta) = \sec(\theta)^2 \tag{6-12b}$$

Do you see what you had to "know" to arrive at the result? You had to use the basic identities in an organized way. Recognizing that the right-hand side reduced to secant squared, you left the first secant on the left alone and rewrote tan as sin over cos. Then you wrote csc as 1 over sin and canceled the sin terms. Last, you converted the cos term in the denominator to sec in the numerator and multiplied it with the first sec term (the one you didn't change). You need to do a few of these problems to understand the process, but most people get little satisfaction from proving identities like these "the hard way."

Your Turn. Prove that $\cos(\theta)\tan(\theta)\sin(\theta) = (\cot(\theta)^2 + 1)^{-1}$

Answer: ______________________________

Example 6-3

Prove that $\dfrac{1+\cos(x)}{\sin(x)} = \dfrac{\sin(x)}{1-\cos(x)}$

Solution:

> e3 := (1+cos(x)) / sin(x) = sin(x) / (1-cos(x));

$$e3 := \frac{1 + \cos(x)}{\sin(x)} = \frac{\sin(x)}{1 - \cos(x)}$$

> simplify(e3);

$$\frac{1 + \cos(x)}{\sin(x)} = -\frac{\sin(x)}{\cos(x) - 1}$$

Maple doesn't do a very good job with trigonometric fractions. On the other hand:

> simplify(lhs(e3)-rhs(e3));

$$0$$

Subtracting the left-hand side of Equation e3 from the right-hand side produces exactly 0 as a result. Therefore, the original equation must be an identity.

Your Turn. Prove that $\frac{1-\sin(y)}{\sin(y)\cot(y)} = \frac{\cos(y)}{1+\sin(y)}$.

Answer: ______________________________

Example 6-4

Show whether $\frac{\sin(x)(\cot(x)-1)}{\sin(x)-\cos(x)} = \cot(x) - 1$ is an identity.

Solution:

```
> e4 := sin(x)*( cot(x)-1)/ ( sin(x)-cos(x) ) = cot(x)-1;
```

$$e4 := \frac{\sin(x)(\cot(x)-1)}{\sin(x)-\cos(x)} = \cot(x) - 1$$

```
> simplify( e4 );
```

$$-1 = \frac{\cos(x)-\sin(x)}{\sin(x)}$$

But how can one side of the supposed identity reduce to 1 and the other reduce to a function of sin and cos? We investigate further by evaluating the right-hand side for some chosen value of x. Let's take an angle of 30°, which is π/6 in radian measure.

```
> evalf(subs( x = Pi/6, e4));
```

$$-1.000000000 = .732050807$$

Well, –1 does *not* equal 0.732! Since we have shown that there is at least one value of x that causes the right-hand side to differ from the left-hand side, Statement e4 is not an identity. In fact, we could ask Maple to solve Statement e4, because it is an equation.

```
> solve(e4, x);
```

$$\frac{1}{2}\pi$$

You can confirm this by substituting this value back into the original equation.

```
> subs( x = Pi/2, e4);
```

$$\frac{\sin\left(\frac{1}{2}\pi\right)\left(\cot\left(\frac{1}{2}\pi\right)-1\right)}{\sin\left(\frac{1}{2}\pi\right)-\cos\left(\frac{1}{2}\pi\right)} = \cot\left(\frac{1}{2}\pi\right) - 1$$

Unfortunately, this seems to be more complicated, but we can attempt a simplification.

> simplify(subs(x = Pi/2, e4));

```
-1 = -1
```

That has solved it. By substituting our solution back into the original equation and simplifying, we have verified the solution. Note that this may be only one of a number of solutions to this equation. If a trigonometric equation has one solution, it often has an infinite number of solutions. Further analysis is required to see how these other solutions are related to the one Maple arrived at in reponse to the *solve* command.

Paper and Pencil Exercises

PP6–1

$$\tan(\theta)^2 + 1 = \sec(\theta)^2 \qquad (6\text{-}10a)$$

$$\cot(\theta)^2 + 1 = \csc(\theta)^2 \qquad (6\text{-}11a)$$

Prove Equations 6-10 and 6-11 by expressing the trigonometric functions as the ratios of the sides in the triangle of Figure 6.1. Simplify the resulting algebraic equation. Hint: the number 1 in these equations can be written as x^2/x^2 in the first equation and as y^2/y^2 in the second equation, which causes both left-hand terms to have a common denominator. Add the numerators and simplify to achieve the desired result.

PP6–2

Prove the following fundamental trigonometric identities:

(a) $\sin(x)\csc(x) = 1$ (use Equation 6-4) *Answer:* ____________________

(b) $1 + \dfrac{1}{\tan^2(x)} = \csc^2(x)$ *Answer:* ____________________

(c) $1 + \dfrac{1}{\tan^2(x)} = \dfrac{1}{\sin^2(x)}$ *Answer:* ____________________

PP6–3

Simplify $\cos(x)\tan^2(x) + \cos(x)$ to a single trigonometric expression:

Answer: ____

PP6–4

Simplify $\frac{1}{\tan(\theta)\csc(\theta)}$ *Answer:* ____________________

PP6–5

Simplify $\frac{\cos(x)\csc(x)}{\tan(x)}$ *Answer:* ____________________

PP6–6

Express $\sin(x) - \cos^2(x)\sin(x)$ as a single trigonometric power.

Answer: ____

PP6–7

Simplify $(1 + \cot^2(x))\sin(x)$ *Answer:* ____________________

PP6–8

Simplify $\cos^2(x)(1 + \tan^2(x))$ *Answer:* ____________________

PP6–9

Simplify $(\tan(x) + \cot(x))\csc(x)$ *Answer:* ____________________

PP6–10

Simplify $\frac{\tan(x)}{\sec(x) - \cos(x)}$ *Answer:* ____________________

Maple Lab

ML6–1: Method 1, Simplification

If the trigonometric expressions are not too complicated, try simplifying each side of the equal sign separately. For example, prove $\frac{\sin(y)}{1 + \cos(y)} = \frac{1 - \cos(y)}{\sin(y)}$ is an identity.

Solution. Simplify each side of the equation:

> **simplify(sin(y) / (1+cos(y))) ;**

$$\frac{\sin(y)}{1 + \cos(y)}$$

There is no change. The simplify command has not done anything with the expression.

> **simplify((1- cos(y)) / sin(y));**

$$\frac{\sin(y)}{1 + \cos(y)}$$

Maple does change this expression; it transforms it to a replica of the first output. Since the left-hand side transforms to the same expression as the right-hand side, the two sides are identical and the equation is an identity.

Try this method on the following examples. State which ones are *not* an identity:

(a) $\tan^2(x)\cos^2(x) + \dfrac{\sin^2(x)}{\tan^2(x)} = 1$ (obviously, the right-hand side is already simplified. Simplify the left-hand side).

> ; (Write the Maple command here)____________________________________

*Answer:*____

(b) $\dfrac{\sin^3(x)\cos^3(x)}{(\cos(x) + \sin(x))} = \sec(x)$

> ; (Write the Maple command here)____________________________________

*Answer:*____

(c) $\dfrac{1 + \tan(x)}{\sin(x)} - \dfrac{1}{\cos(x)} = \dfrac{1}{\sin(x)}$

> ; (Write the Maple command here)____________________________________

*Answer:*____

(d) $\sec(x)\left(\sec(x) - \dfrac{1}{\sec(x)}\right) + \dfrac{\cos(x) - \sin(x)}{\cos(x)} = \sec^2(x) - \tan(x)$

> ; (Write the Maple command here)____________________________________

*Answer:*____

(e) $\dfrac{\cos(x)+\sin(x)}{\cos(x)-\sin(x)} = \dfrac{1-\tan(x)}{1+\tan(x)}$

> ; (Write the Maple command here) ______________________________

Answer: ______________________________

(f) $\dfrac{2}{\sin(x)} = \dfrac{\sin(x)}{1+\cos(x)} + \dfrac{1+\cos(x)}{\sin(x)}$ ______________________________

> ; (Write the Maple command here) ______________________________

Answer: ______________________________

____

(g) $\tan(x) - \csc(x)\sec(x)(1-2\cos^2(x)) = \dfrac{\tan(x)-\sec(x)-1}{\tan(x)-\sec(x)+1} - \dfrac{1-2\cos^2(x)}{\sin(x)\cos(x)} - \sec(x)$

> ; (Write the Maple command here) ______________________________

Answer: ______________________________

ML6–2: Method 2, Give Maple Some Help

Sometimes Maple refuses to simplify trigonometric expressions to the form you want. For instance, try:

> e62 := simplify(1 / (1-cos(x)^2));

$$e62 := -\frac{1}{-1+\cos(x)^2}$$

Maple has changed a couple of signs, but it hasn't made the obvious simplification: $1-\cos^2(x) = \sin^2(x)$.

You can give Maple some help by inputting the trigonometric identity you want it to apply. Define t as

> t := sin(x)^2 + cos(x)^2 = 1;

$$t := \sin(x)^2 + \cos(x)^2 = 1$$

Now try the simplify command again, but this time specify that you want Maple to use t as part of its simplification procedure. (Note that you place t inside curly brackets.)

> simplify(e62, {t});

$$\frac{1}{\sin(x)^2}$$

The simplification has been performed in just the way you wanted.

Is ML6–1(g) an identity? Apply this method to see if you get a better simplification.

> ; (Write the Maple command here) ______________________________

Answer: ______________________________

ML6–3: Method 3, Rule Out False Cases by Evaluation

Most trigonometric equations are not identities. You can rule out these cases by evaluating the two sides of the equation for some randomly chosen angles. If the sides are the same, you have found a *solution to the equation.* If the two sides are the same for a few scattered choices of the angle, the equation is likely an identity, but you must use some means like method 1 to be sure.

If the two sides evaluate to different numbers for even one choice of angle, the equation *cannot* be an identity. To use this method, it's important to type the equation into Maple and give it a name. Then you can work with the equation, or parts of the equation, in many different ways.

(a) Prove that $\frac{1}{\sec(\theta)\tan(\theta)} = \sec(\theta) - \cot(\theta)$ is *not* an identity. Give the equation a name:

> eq63 := 1/ (sec(theta)*tan(theta)) = sec(theta)-cot(theta);

$$eq63 := \frac{1}{\sec(\theta)\tan(\theta)} = \sec(\theta) - \cot(\theta)$$

Evaluate this equation for some values of θ. Show one of those values for which the right-hand side (rhs) is not equal to the left-hand side (lhs).

> ; (Write your Maple commands here) ______________________________

Answer: At θ = __________, lhs = __________, and rhs = __________ .

(b) Prove that $\frac{\sec(x) + \csc(x)}{\tan(x) + \cot(x)} = \sin(x) - \cos(x)$ is not an identity.

> ; (Write your Maple commands here) ______________________________

Answer: At θ = __________, lhs = __________, and rhs = __________ .

ML6–4: Method 4, Rule Out False Cases by Plotting

If you plot the right-hand and left-hand sides of an equation together in the same graph and find two different curves, then the original equation is certainly not an identity.

(a) Prove that $\dfrac{\cos(\theta) + \sin(\theta)}{1 + \tan(\theta)} = \sin(\theta)$ is not an identity:

> ; (Write your Maple commands, including the *plot* command, here) ______________

____

Answer: At θ = __________, lhs = __________, and rhs = __________ .

(b) Does the plot for $\sin(x)\dfrac{\sin(x)}{1 - \cos(x)} + \dfrac{1 - \cos(x)}{\sin(x)} = 2$ supply strong evidence that it is an identity?

> ; (Write your Maple commands, including the *plot* command, here) ______________

____

Answer: ____

ML6–5: More on Plotting Trigonometric Equations

(a) Plot the two sides of $\sqrt{\dfrac{1 + \cos(\theta)}{1 - \cos(\theta)}} = 1 + \cos(\theta)$. Look at the form of the equation. There is a $1 + \cos(\theta)$ term on both sides. Does this equation look like a possible identity to you?

Let's see if the graph of the two sides provides strong evidence that the equation is an identity. Construct such a graph and examine the plot carefully. What does the graph allow you to say about the two sides, especially for the range, $x = 0.8 .. 5.4$?

> ; (Write your Maple commands, including the *plot* command, here) ______________

____

Answer: ____

(b) Now consider the equation $\sqrt{\dfrac{1 + \cos(\theta)}{1 - \cos(\theta)}}\sin(\theta) = 1 + \cos(\theta)$. Plot both sides for $\theta = 0 .. 2\pi$. What do you conclude? (Make sure to discuss the square root term and its two possible values.)

> ; (Write your Maple commands, including the *plot* command, here) ________________

____

Answer: ____

____

____

ML6–6: Simplifying Trigonometric Expressions Containing Square Roots

Maple assumes that it is working in a domain that includes complex numbers as well as real numbers. This means that you must take extra care when working with terms containing square roots. For example, in this problem you want to work only with real numbers and positive square roots.

Investigate the problem of simplifying:

> **e66 := sqrt(1 + 2 * sin(x) + sin(x)^2);**

$$e66 := \sqrt{1 + 2\sin(x) + \sin(x)^2}$$

(a) What is the result of using *simplify* on statement e66?

> ; (Write the Maple command and its output) ________________________________

(b) Factor the expression inside the square root sign of statement e66. State the result.

Answer: ____

(c) Teach Maple to get the same answer. Use the commands *abs*, *factor*, *simplify*, and *subs*, as well as other commands you might think of, including your own definition of the square root function, such as:

> **sr := x -> sqrt(abs(x));** (This function makes sure you are taking the square root of a positive number.)

Write your solution:

Answer: ____

____

____

____

____

ML6–7

In Maple, you must make a distinction between "the name of a function" and "a function evaluated at a given value of its variable."

> **sin ;** (This is the *name* of a function)

> **sin(x) ;** (This is the sine function, evaluated at a given value [namely, x])

(a) What will these commands produce? Write your answer before trying them out. If Maple's output is different from your prediction, you need to review the distinction between a *name* and a *value*.

> **sin * Pi; evalf(sin * Pi);** *Answer:* ____________________

(b) Define "the squaring function" as:

> **sq := x -> x^2;**

sq := $x \rightarrow x^2$

What do these commands produce?

> **sq x ; sq * x ; x sq ; sq(x) ; sq(9);** *Answer:* sq x: ____________

Answer: sq $*x$: ____________

Answer: x sq : ____________

Answer: sq(x) : ____________

Answer: sq(9) : ____________

(c) Say you prefer the notation $\sin^2(x)$, as given in most textbooks, for the square of $\sin(x)$. If so, you will want to define a function sin2(x) to do the job. Write the Maple command that defines this function:

Answer: ____

ML6–8

(a) What is the result of:

> **simplify(2-2 * sin(x)^2);** *Answer:* ____________________

(b) State the result of the command:

> **simplify((1-cos(x)^2) / sin(x)^2);** *Answer:* ____________________

(c) State the result of the command, but first predict the result:

```
> simplify( sqrt (x^2 ) );
```

Answer: ____________________

Explorations

E6-1

When they are first introduced, the calculation of trigonometric identities may seem like a silly exercise. After all, you may wonder, where will you encounter a need to apply this knowledge in any of your other subjects? However, since trigonometry is used to describe electrical oscillations and mechanical vibrations, the need to deal with trigonometric formulas is certainly encountered in electrical and mechanical technology, as well as in other branches of science and technology.

In many cases, two sine waves are added. Is the sum also a sine wave? (Note that a cosine function is considered to be a sine wave in this discussion.) The general term for a wave like this is a *sinusoid.*

Examine the following cases:

(a) $A\sin(\omega t) + B\cos(\omega t)$

(b) $A\sin(\omega t) + B\sin(\omega t)$

(c) $A\sin(\omega t) + B\sin(\omega t + \alpha)$

Plot the expressions in (a), (b), and (c) using the values $A = 3$, $B = 4$, and $\omega = 2$ over the interval $t = 0$ .. Pi.

Do any (or all) these sums resemble a sinusoid?

CHAPTER 7

Vectors and Trigonometry

Objectives for This Chapter

1. Know the definition of a vector in Maple
2. Use the components of a vector as a means of solving vector problems
3. Know the methods of addition and subtraction of vectors
4. Examine applications of scalar multiplication
5. Define the scalar product (dot product) and the vector product (cross product)
6. Use the dot product and the cross product to derive the cosine and sine laws

Maple Commands Used in This Chapter

evalf(angle(C,A));	Find the angle between two vectors (usually requires *evalf*).
evalm(A + B)	Evaluate a vector result (similar to *evalf*).
norm(A, 2)	Calculate the magnitude of a vector from its components.
with(linalg)	Load the package that contains vector functions.
vector([Ax, Ay])	Define a vector with components *Ax* and *Ay*.

linalg[dotprod](A, B)   Calculate the dot product of vectors A and B without loading *linalg*.

The Concept of a Vector

John Wallis (1616–1703) was the first to attempt to represent the "imaginary" solutions of a quadratic equation. He chose to measure the real part of the solution along the x axis and the imaginary part along the y axis. This bold step was neglected for a century, when it was again popularized by the Norwegian surveyor Caspar Wessel (1745–1818) and the French mathematician J. R. Argand. The construction is accomplished by starting at the origin of an (x, y) coordinate system and going out a distance, x, along the real axis, and then up a distance, y, in a direction perpendicular to x (Figure 7.1).

It took a while longer for mathematicians to realize that vectors could be defined without relating them to complex numbers. For example, the vectors that we will define in this chapter contain only real numbers. They are useful because they express powerful concepts in a simple notation. They are used extensively in physics, mechanics, and engineering.

The *vector* $\overrightarrow{\mathbf{A}}$ is defined as the *directed line* $\overrightarrow{OP}$ that joins the origin, O, to the point $P(x, y)$.

You can see from this definition that the vector $\overrightarrow{\mathbf{A}}$ is nothing more than the familiar hypotenuse of the right-angled triangle containing x and y, with a sense of direction included.

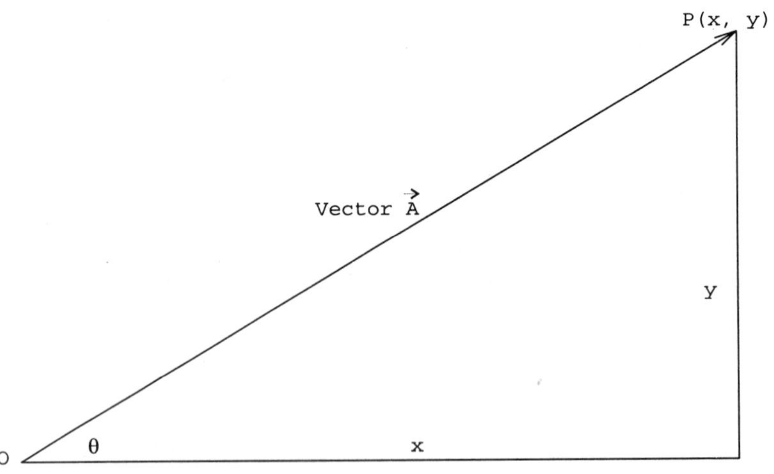

Figure 7.1 The Definition of a Vector

You think of the line *OP* as going from *O* to *P*. (You will see shortly that the line going the other direction, from *P* to *O*, is called $-\vec{\mathbf{A}}$.)

Vectors are especially useful in physics, where they represent directed quantities, like force or displacement, in a natural way. Since vectors are so closely related to trigonometry, they will be discussed here.

We need some terminology to be able to talk about the parts of a vector. The *initial point*, or the point where the vector begins, is the point *O*. It is also called the *tail* of the vector. The *terminal point* is the point where the vector ends. This point is called the *tip*. The tip of a vector is drawn with a little arrowhead so that the direction of the vector is made clear. The direction of a vector goes *from* the tail *to* the tip (refer to Figure 7.1). The *components* of a vector are the lengths x and y. The angle the vector makes with the x axis is θ. The *magnitude* of a vector is the length of the hypotenuse, independent of its direction. The magnitude of $\vec{\mathbf{A}}$ is denoted by $|\vec{\mathbf{A}}|$. We can relate the magnitude of the vector $|\vec{\mathbf{A}}|$ to the magnitude of its horizontal and vertical components using the basic trigonometric definitions of sine and cosine.

$$x = |\vec{\mathbf{A}}|\cos(\theta) \text{ (the } x \text{ component of the vector } \vec{\mathbf{A}}\text{)} \tag{7-1a}$$

$$y = |\vec{\mathbf{A}}|\sin(\theta) \text{ (the } y \text{ component of the vector } \vec{\mathbf{A}}\text{)} \tag{7-1b}$$

$$|\vec{\mathbf{A}}| = \sqrt{x^2 + y^2} \tag{7-1c}$$

Addition of Two Vectors

The addition of two vectors is best thought of as "tail to tip" addition. You place the tail of the first vector, $\vec{\mathbf{A}}$, at the origin, *O*. Then you place the tail of vector $\vec{\mathbf{B}}$ on the tip of vector $\vec{\mathbf{A}}$, maintaining its proper angle with the x axis. The sum $\vec{\mathbf{A}} + \vec{\mathbf{B}} = \vec{\mathbf{C}}$ is the line from *O* to P_2. Figure 7.2 shows the geometrical construction for the sum of two vectors.

The diagram shows that it is easy to add two vectors mathematically if you know their components. Let x_3 and y_3 be the components of $\vec{\mathbf{C}}$. In that case, $x_3 = x_1 + x_2$ and $y_3 = y_1 + y_2$. Once you know the components of a vector, it is easy to reassemble it. Again we apply the basic trigonometric definitions:

$$|\vec{\mathbf{C}}| = \sqrt{x_3^2 + y_3^2} = \sqrt{(x_1^2 + x_2^2) + (y_1^2 + y_2^2)} \tag{7-2a}$$

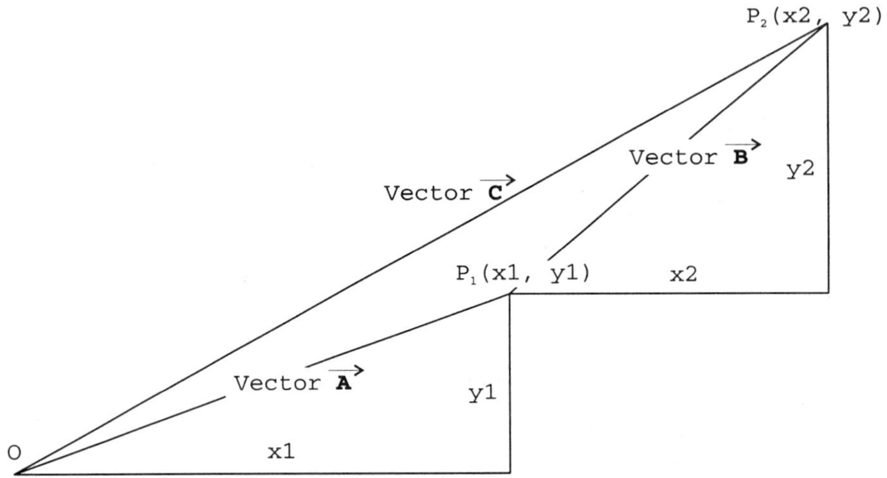

Figure 7.2 The Sum of Two Vectors

The angle θ that $\vec{\mathbf{C}}$ makes with the x axis is found from the equation:

$$\tan(\theta) = \frac{y_3}{x_3} = \frac{y_1 + y_2}{x_1 + x_2} \tag{7-2b}$$

A vector can be specified in terms of *unit vectors*. The vector **i** is the unit vector in the x direction, and the vector **j** is the unit vector in the y direction. The vector $x\mathbf{i}$ is in the same direction as **i**, and is x times as long. The number x is called a *scalar*. The process of multiplying a vector by a real number is called *scalar multiplication*. Thus, if x_1 and y_1 are the x and y components of a vector **A**, then $\mathbf{A} = \mathbf{i}x_1 + \mathbf{j}y_1$. (From now on in this chapter, we will denote vectors with bold letters, omitting the little arrow above the letter.) Any vector can be written as a sum of its x and y components. This is especially important because it gives us a simple way of adding two vectors. Realize that two vectors which both lie along the x axis will have a sum which also lies along the x axis. The magnitude of the *resultant* will be just the sum of the individual magnitudes. Let the two quantities be called **p** and **q**. Since these vectors lie along the x axis, they can be represented as $\mathbf{p} = \mathbf{i}p$ and $\mathbf{q} = \mathbf{i}q$. Consequently, $\mathbf{p} + \mathbf{q} = \mathbf{i}p + \mathbf{i}q = \mathbf{i}(p + q)$.

Since vectors that are oriented in the same direction can be added just like numbers, we can add any two vectors **A** and **B** by adding together the x components, and then the y components. Setting $A_x = x_1$, $A_y = y_1$, $B_x = x_2$, $B_y = y_2$, we have:

$$\mathbf{C} = \mathbf{A} + \mathbf{B} = A_x\mathbf{i} + A_y\mathbf{j} + B_x\mathbf{i} + B_y\mathbf{j} = (A_x + B_x)\mathbf{i} + (A_y + B_y)\mathbf{j}. \tag{7-3}$$

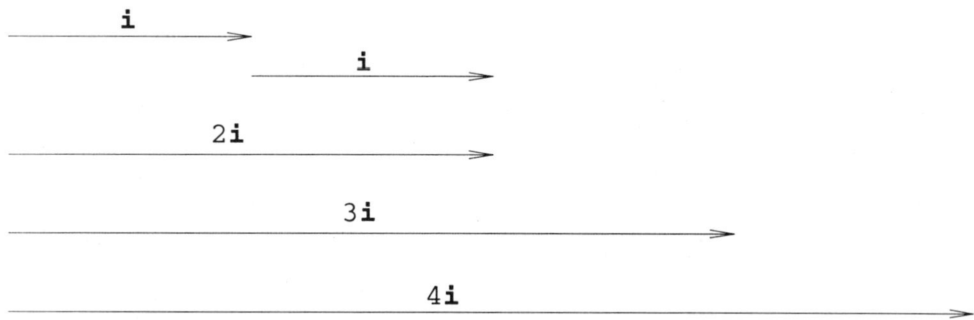

Figure 7.3 Adding Unidirectional Vectors

Figure 7.3 shows the unit vector **i** and some of its scalar multiples. The unit vector **i** is shown twice to remind you that a vector can be moved from the origin and still be considered the same vector. The starting point for a vector's tail is immaterial: only its length and direction are important. Scalar multiples of **i** are all in the same direction, which is that of **i** itself.

Vectors can be subtracted by adding the negative of a vector. The vector difference $\mathbf{A} - \mathbf{B}$ is defined as $\mathbf{A} + (-\mathbf{B})$. The negative of a vector is found simply by reversing its direction. Interchange the tip and tail, and you have the negative of the original vector.

The Dot Product of Two Vectors

You have seen that vectors can be added and subtracted, but it is not obvious how the multiplication of two vectors should be defined. It has been found useful to define two different ways to multiply vectors. The first of these is called the *dot product*. The dot product is written as $\mathbf{A} \cdot \mathbf{B}$,

$$\mathbf{A} \cdot \mathbf{B} = |A||B|\cos(\theta), \tag{7-4}$$

where θ is the angle between **A** and **B**.

Figure 7.4 shows the diagram for the dot product of two vectors, **A** and **B**. The dot product is a *scalar* quantity. This way of multiplying two vectors produces a number, not a vector. The reason for this definition is its usefulness. Many physical occurrences can be described in terms of the dot product. The most commonly known of these is the action of a force on a moving object. The *work done* by the force is equal to the dot product of the force and the distance moved.

$$W = \mathbf{F} \cdot \mathbf{x} = F\,x\cos(\theta) \tag{7-5}$$

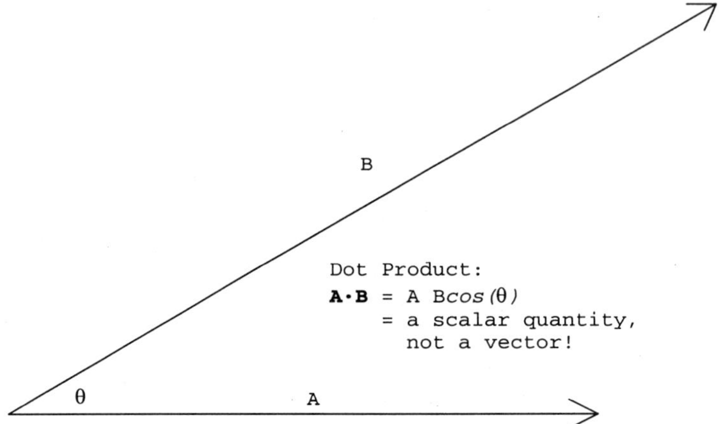

Figure 7.4 The Dot Product

The dot product formula is easy to work with if the vectors are expressed in component form. The reason is the exceptionally simple rule for taking the dot product of the unit vectors **i** and **j**.

$$\mathbf{i}\cdot\mathbf{i} = \mathbf{j}\cdot\mathbf{j} = 1,\ \mathbf{i}\cdot\mathbf{j} = \mathbf{j}\cdot\mathbf{i} = 0 \qquad (7\text{-}6)$$

These results follow directly from the definition of the dot product. If two vectors are in the same direction, like **i**·**i**, the angle between them is 0. The cosine of 0 is 1, so the magnitudes of the vectors are multiplied. If the vectors are at right angles to one another, like **i** and **j**, the angle between them is 90°, and the cos(90) = 0. Equation 7-7 follows immediately. If **A** and **B** are any two vectors, we express them in component form to find the dot product.

$$\begin{aligned}\mathbf{A}\cdot\mathbf{B} &= (A_x\mathbf{i} + A_y\mathbf{j})\cdot(B_x\mathbf{i} + B_y\mathbf{j})\\ &= A_xB_x\mathbf{i}\cdot\mathbf{i} + A_xB_y\mathbf{i}\cdot\mathbf{j} + A_yB_x\mathbf{j}\cdot\mathbf{i} + A_xB_x\mathbf{j}\cdot\mathbf{j}\\ &= A_xB_x + A_yB_y \qquad (7\text{-}7)\end{aligned}$$

The vector dot product formula can be generalized to many dimensions. Since the Maple routines for vector manipulation are designed to include the general case, this imposes a bit of complexity.

Maple uses the *linalg* package to define a large number of vector operations. In fact, the *linalg* package in Maple defines many more operations than we have use for in this book. If the machine you are working on has lots of memory, you can load the entire package. (The command to load the *linalg* package follows.) After loading it, you can add and subtract two vectors, and you can calculate dot and cross products. If you are short of memory, you can still use the commands you need. All it costs is a little more typing. For example, to calculate the dot product of two vectors, you would define them as shown, and then issue the command, *linalg[dotprod](A, B);*. Only the *dotprod* routine is loaded, not the entire *linalg* package.

> with (linalg):

> A := vector([Ax,Ay]);

$$A := [\, Ax \;\; Ay] \tag{7–8}$$

> B := vector([Bx, By]);

$$B := [\, Bx \;\; By] \tag{7–9}$$

> linalg[dotprod](A, B);

$$AxBx + AyBy \tag{7-10}$$

Example 7-1: Some Vector Operations

If $\vec{\mathbf{A}} = 25\mathbf{i} + 35\mathbf{j}$, and $\vec{\mathbf{B}} = 15\mathbf{i} - 10\mathbf{j}$, find $\vec{\mathbf{A}} + \vec{\mathbf{B}}$, $\vec{\mathbf{A}} - \vec{\mathbf{B}}$, and $\vec{\mathbf{A}} \cdot \vec{\mathbf{B}}$.

Solution. Define the vectors in Maple after loading the *linalg* package.

> with(linalg);

Warning, new definition for norm
Warning, new definition for trace

>A := vector([25, 35]); B := vector([15, -10]);

```
A := vector([ 25, 35] )
B := vector([ 15, -10] )
```

> evalm(A+B);

```
[ 40, 25]
```

> evalm(A-B);

```
[ 10, 45]
```

> dotprod(A, B);

```
25
```

Example 7-2: Vector Addition When the Magnitude and Angle Are Given

We will use the notation **B** at θ to indicate the magnitude and direction of a vector. This is equivalent to giving its *x* and *y* components, since one form can be transformed into the other using the trigonometry of the right-angled triangle. When we use this notation, θ is the angle the vector makes with the x axis. Thus, a vector whose components are **B** = [1, 1] can be written as **B** = $\sqrt{2}$ at 45°.

If **A** = 10 at 30° and **B** = 15 at 45°, find **A** + **B**, **A** – **B**, and **A** · **B**.

Convert the vectors to rectangular form as follows:

> A := vector([10*Cos(30), 10*Sin(30)]); B := vector([15*Cos(45), 15*Sin(45)]);

```
A := [ 8.660254040, 5.000000000]
B := [ 10.60660172, 10.60660172]
```

> evalm(A + B); evalm(A-B); dotprod(A, B);

```
[ 19.26685576, 15.60660172]
[ -1.946347680, -5.606601720]
144.8888740
```

Note that the definition of Sin, Cos, and Tan that we are using here converts all angles to decimal approximations. We could have used a definition of these functions that gave exact values for simple angles had we wished.

Your Turn. Given **A** = [17.5, –33.2] and **B** = [–17.5, 33.2], find **A** + **B**, **A** – **B**, and **A** · **B**.

A + **B** = ____

A – **B** = ____

A · **B** = ____

Example 7-3

The angle between vectors **A** and **B** is 20°, and the magnitudes of **A** and **B** are 135 and 225, respectively. Vector **A** lies along the positive x axis and **B** lies above **A**. Find **A** + **B**, **A** – **B** and **A** · **B**.

Solution. Given the orientations of the two vectors, we can assign components to each vector. Since **A** lies on the x axis, its y component is 0.

> A := vector([135, 0]); B := vector([225*Cos(20), 225*Sin(20)]);

```
A := [ 135, 0]
B := [ 211.4308397, 76.95453224]
```

> evalm(A + B); evalm(A-B); dotprod(A, B);

```
[ 346.4308397, 76.95453224]
[ -76.4308397, -76.95453224]
28543.16336
```

Your Turn. Given **A** = 720 at 15° and **B** = 490 at –15°, find **A + B, A – B,** and **A · B.**

A + B = ______________________________

A – B = ______________________________

A · B = ______________________________

The Cosine Law

Once the dot product has been defined and its properties understood, it is easy to prove a generalization of the Pythagorean theorem. Let a triangle have sides *a*, *b*, and *c*. You must have a right-angled triangle if you want to use the formula,

$$a^2 + b^2 = c^2 \tag{7-11}$$

but what if the triangle contains no right angle, such as the one shown in Figure 7.5?

Let θ be the exterior angle between sides **A** and **B**. Angle θ is therefore the angle between **A** and **B**, so we can apply the dot product formula directly. Express the three sides of the triangle as vectors **A**, **B**, and **C**. Thus, we have the vector relationship:

$$\mathbf{C} = \mathbf{A} + \mathbf{B} \tag{7-12}$$

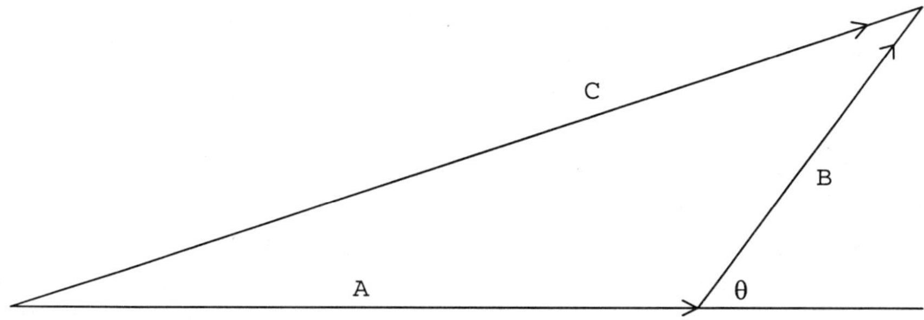

Figure 7.5 Diagram for Derivation of the Cosine Law

Take the dot product of **C** with itself and apply the rules for the vector manipulation of **A** + **B**.

$$\mathbf{C} \cdot \mathbf{C} = (\mathbf{A} + \mathbf{B}) \cdot (\mathbf{A} + \mathbf{B})$$

$$C^2 = \mathbf{A} \cdot \mathbf{A} + \mathbf{B} \cdot \mathbf{B} + \mathbf{B} \cdot \mathbf{A} + \mathbf{A} \cdot \mathbf{B}$$

$$C^2 = A^2 + B^2 + 2AB\ cos(\theta) \qquad (7\text{-}13)$$

Equation 7-13 is called the *cosine law.* Note that the angle θ is the exterior angle of the triangle and also the angle between **A** and **B**, as required by the definition of the dot product. This formula allows you to find the third side of a triangle if you know two sides and the included angle—the angle between the sides. The exterior angle is found by subtracting the included angle from 180° (or 2π, if you are working in radians).

Example 7-4

Use Figure 7.5 with $A = 45.5$, $B = 27.3$, and $\theta = 1$ radian to find side C.

Solution. **C** = **A** + **B** and

> with (linalg);

> A := vector([45, 0]); B := vector([27.3*cos(1), 27.3*sin(1)]);

```
A := [ 45, 0]
B := [ 27.3 cos(1), 27.3 sin(1)]
```

> C := evalm(A + B); (Since A and B are vectors and they sum to C, this is all we have to do.)

```
C := [ 59.75025295, 22.97215789]
```

Compare the trigonometric solution: $C^2 = A^2 + B^2 + 2Ab\ \cos(\theta) = 45.5^2 + 27.3^2 + 2(45.5)(27.3)\cos(1)$. This method produces a value for C of 64. How do we compare the value of C we found using Maple to this one? In other words, how do we convert C in rectangular form to a magnitude? One way is to use the *norm* command. The form of the command we must use is:

> norm(C, 2);

```
64.01416067
```

Note that it takes further work to find the angle that C makes with the x axis. The vector solution carries with it information about the angle, since the x and y components are part of the solution. When you use the vector solution, finding the magnitude of C requires further work. There are trade-offs with every approach. That's why it is wise to know more than one method.

You could use:

> evalf(angle(C,A));

```
.3670468592
```

to find the angle between *C* and *A*. As usual, *evalf* converts the answer to decimal form.

Example 7-5

Given the vectors **A** = [16, 7] and **B** = [2, 11], find the angle θ between them.

Solution. The dot product of two vectors involves the cosine of the angle between them. We will use Equation 7-4. Our task is to express the magnitudes of **A** and **B** using Maple commands along with the dot product. We must first define the vectors in Maple, then form the command that computes the angle in degrees. Do you see how we used the "function of a function" approach to arrive at the final command?

> with (linalg);

> A := vector([16, 7]): B := vector([2, 11]):

> theta = evalf(180/Pi*arccos(dotprod(A, B)/(norm(A, 2)*norm(B, 2))));

```
θ = 56.06577576
```

We will check the answer by using the Maple command for finding the angle between two vectors.

> theta = evalf(angle(A, B)*180/Pi);

```
θ = 56.06577578
```

Notice that the answers differ in the last decimal place. Two different approaches produced two slightly different answers. This is typical in numerical work.

Your Turn: Use the definition of the cross product to calculate the sine of the angle between **A** and **B**, and then look up the arcsine to find θ.

Answer: Maple command:____

Angle θ: ____

Does your answer check with the results of Example 7-5?

Answer: ____

The Vector Cross Product

We are interested only in some selected facts about the vector cross product. It is really a topic for linear algebra, so we will be content to define it and then use the definition to derive a rule for solving triangles based on the sine function. Vectors are three-dimensional objects, but we will confine ourselves to the x, y plane, since this is the surface on which trigonometry is done. The vector cross product will require that we consider the z axis as well, but only in a superficial way. This method of vector multiplication has a vector as a result. The cross product of a vector **A** with a vector **B** yields a vector **C** whose magnitude is $AB\sin(\theta)$, where θ is the angle that rotates **A** into **B** in the most efficient possible way (that is, through the fewest number of degrees). The definition of the vector cross product is more tricky than that of the dot product formula and needs more explanation. The cross product is written as

$$\mathbf{C} = \mathbf{A} \times \mathbf{B}$$

$$|\mathbf{C}| = |\mathbf{A}||\mathbf{B}| \sin(\theta) \tag{7-14}$$

The first part of Equation 7-14 shows how the cross product is represented, and the second part shows how to calculate the magnitude of **C**. We will confine the vectors **A** and **B** to the x, y plane. In this case, the direction of **C** is along the z axis. The direction for the vector cross product is defined as that perpendicular to the plane containing **A** and **B**. Furthermore, the direction of **C** is given by the *right hand rule*: using your right hand, place your fingers in the direction of **A** and then curve them around to the direction of **B**. As you perform this rotation of **A** into **B**, your thumb, if held straight out from your fingers, will point in the direction of the vector cross product, **C**. A diagram helps. Figure 7.6 shows two vectors, **A** = 5**i** + **j** and **B** = 2**i** + 2**j**. The cross product can be found by finding the angle these vectors make with one another. One way of determining the angle is to take the dot product, which allows us to find the cosine of the angle between **A** and **B**.

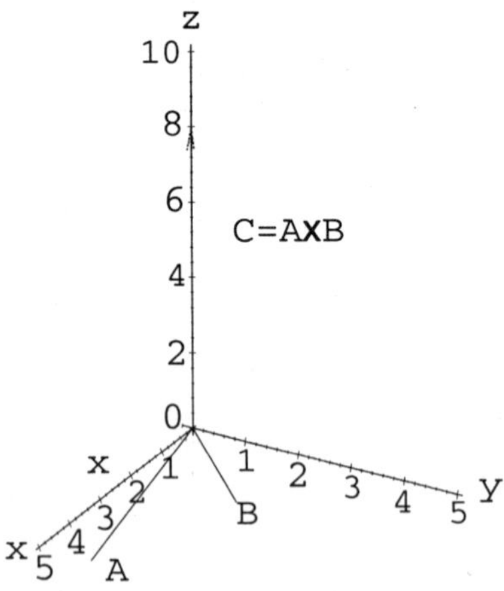

Figure 7.6 The Cross Product of Two Vectors, **A** and **B**

$$\mathbf{A}\cdot\mathbf{B} = |\boldsymbol{A}||\boldsymbol{B}|\cos(\theta)$$

$$\cos(\theta) = \frac{\boldsymbol{A}\cdot\boldsymbol{B}}{|\boldsymbol{A}||\boldsymbol{B}|} = \frac{(5\boldsymbol{i}+\boldsymbol{j})\cdot(2\boldsymbol{i}+2\boldsymbol{j})}{\sqrt{5^2+1^2}\sqrt{2^2+2^2}}$$

$$\cos(\theta) = \frac{12}{\sqrt{26}\sqrt{8}} = \frac{3\cdot 2\cdot 2}{\sqrt{2}\sqrt{13}\,2\sqrt{2}} = \frac{3}{\sqrt{13}} \tag{7-15}$$

Now that we know what $\cos(\theta)$ is, we find $\sin(\theta)$ by constructing the right triangle based on the cosine values. The construction shows us that $\sin(\theta)$ is $2/\sqrt{13}$, so we can now compute the vector cross product (see Figure 7.7).

$$\mathbf{A}\times\mathbf{B} = |\boldsymbol{A}||\boldsymbol{B}|\sin(\theta)\mathbf{k}$$

$$\mathbf{A}\times\mathbf{B} = \sqrt{13}\sqrt{2}\,2\sqrt{2}\cdot\frac{2}{\sqrt{13}}\mathbf{k}$$

$$\mathbf{A}\times\mathbf{B} = \mathbf{C} = 8\mathbf{k}$$

We have inserted the unit vector **k** in our calculations, since we know that the direction of **C** will be along the *z* axis. We can calculate the value of **C** much more easily by applying the rules for the cross products of the unit vectors **i**, **j**, and **k**. These vectors are all at right angles to one another, so in this case, $\sin(\theta) = 1$. The direction of the cross products are given by the right hand rule. Thus, we have the equations

$$\mathbf{i}\times\mathbf{j} = \mathbf{k},\ \mathbf{j}\times\mathbf{i} = -\mathbf{k}$$

$$\mathbf{j}\times\mathbf{k} = \mathbf{i},\ \mathbf{k}\times\mathbf{j} = -\mathbf{i}$$

$$\mathbf{k}\times\mathbf{i} = \mathbf{j},\ \mathbf{i}\times\mathbf{k} = -\mathbf{j} \tag{7-16}$$

You can verify Equation 7-16 mentally by imagining rotating your right hand from the first-mentioned vector to the second. Your thumb will always point in the direction of the result.

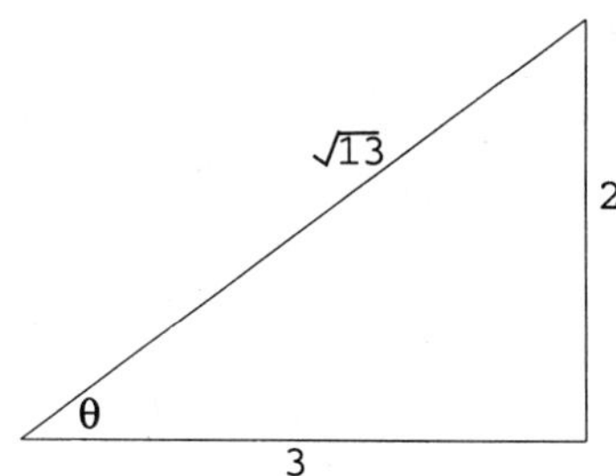

Figure 7.7 Finding sin(θ) when cos(θ) is known

Now that we know how to find the cross product of the unit vectors, we can find **C** by multiplying out the components of **A** and **B**.

$$\begin{aligned} \mathbf{A} \times \mathbf{B} &= (5\mathbf{i} + \mathbf{j}) \times (2\mathbf{i} + 2\mathbf{j}) \\ &= 10\mathbf{i} \times \mathbf{i} + 10\mathbf{i} \times \mathbf{j} + 2\mathbf{j} \times \mathbf{i} + 2\mathbf{j} \times \mathbf{j} \\ &= 10\mathbf{k} - 2\mathbf{k} \\ &= 8\mathbf{k} \end{aligned} \tag{7-17}$$

You must be careful to maintain the order of the vectors as you perform the expansion. Since $\mathbf{i} \times \mathbf{j} = -\mathbf{j} \times \mathbf{i}$, interchanging the order introduces a minus sign. This may be the first time you have encountered a rule for multiplication where you get a different answer if you interchange the terms. We obtain the same result in Equation 7-17 as we got in Equation 7-15. The two methods are equivalent, even though they involve very different calculations.

The Sine Law

We can apply the rule for the vector cross product to three vectors that form a triangle, just as we did to find the cosine law. This time, we will take the cross product instead of the dot product.

The triangle in Figure 7.8 can be represented by $\mathbf{C} = \mathbf{A} + \mathbf{B}$. We will follow tradition in the way the diagram is labeled. The sides are designated by capital letters, and the angles opposite the sides are given by the same lower-case letter. We took the dot product of **C** with itself to derive the cosine law. What happens if we try this with the cross product? The angle between **C** and itself is 0, so the sine is 0 also. *The cross product of a vector with itself is always 0*. We must try something else if we want to get an equation other than 0 = 0!

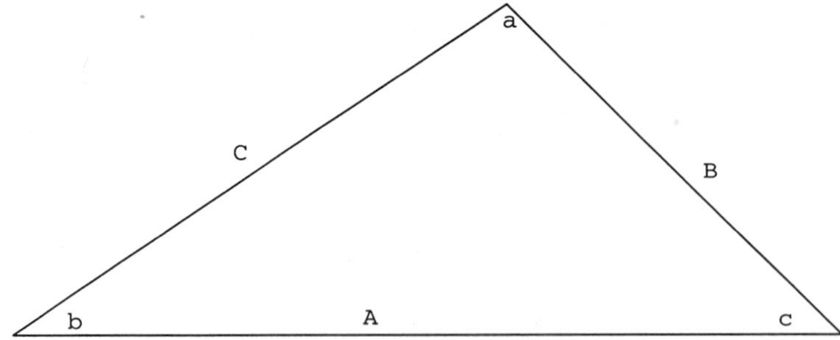

Figure 7.8 Derivation of the Sine Law

Let us write $\mathbf{C} = \mathbf{A} + \mathbf{B}$ as before, but this time take the cross product of **A** with **C**:

$$\mathbf{C} = \mathbf{A} + \mathbf{B}$$

$$\mathbf{A} \times \mathbf{C} = \mathbf{A} \times (\mathbf{A} + \mathbf{B}) = \mathbf{A} \times \mathbf{A} + \mathbf{A} \times \mathbf{B}$$

$$|\boldsymbol{A}||\boldsymbol{C}|\sin(b) = 0 + |\boldsymbol{A}||\boldsymbol{B}|\sin(\pi - c)$$

$$AC\sin(b) = AB\sin(c)$$

$$C\sin(b) = B\sin(c)$$

$$\frac{B}{\sin(b)} = \frac{C}{\sin(c)} \tag{7-18}$$

We have the rather surprising result, in line 2 of Equation 7-18, that the cross product of one side (in this case, A) with either of the other two sides yields the same answer. From that, we deduce a rather simple result in line 6. We have denoted the magnitude of **A** by |**A**| and by A. Since A appears on both sides of line 4 of Equation 7-18, we can eliminate it. Also, we have used the fact that $\sin(\pi - c) = \sin(c)$.

It doesn't matter which two sides we choose for the product. As long as we are careful about our angles, we could write:

$$\mathbf{B} \times \mathbf{C} = \mathbf{B} \times (\mathbf{A} + \mathbf{B})$$

$$BC\sin(2\pi - a) = BA\sin(\pi + c)$$

$$-C\sin(a) = -A\sin(c)$$

$$\frac{A}{\sin(a)} = \frac{C}{\sin(c)} \tag{7-19}$$

Taken together, the last lines of Equations 7-18 and 7-19 give the *law of sines for any triangle.*

$$\frac{A}{\sin(a)} = \frac{B}{\sin(b)} = \frac{C}{\sin(c)} \tag{7-20}$$

Example 7–6

Find all the unknown sides and angles in the triangle of Figure 7.8 if $A = 9.65$ centimeters (cm), $\angle a = 100°$, $B = 5.50$ cm, and $\angle b = 34.0°$.

Solution. We use Equation 7-20 after we have calculated $\angle c$. Since the angles in a triangle add to 180°, $\angle c = (180 - (100 + 34.0)) = 46°$.

---

*Note: A shorthand way of writing angle a is $\angle a$.

$$\frac{A}{\sin(a)} = \frac{B}{\sin(b)}$$

$$\frac{9.65}{\sin(100)} = \frac{B}{\sin(34)}$$

$$B = \frac{(9.65)(0.5592)}{0.9848} = 5.48 \text{ cm}$$

Side C can be found by the same technique. Do the calculation yourself and show that $C =$ 7.05 cm.

The solution to this problem is more cumbersome if you tried to base it on the definition of the cross product. The components of two of the vectors that make up the cross product are not known initially, thus requiring extra calculations.

Your Turn: Find all the unknown sides and angles in Figure 7.8 if:

(a) $A = 10.5$ cm, $\angle a = 97°$, $\angle b = 33°$. *Answer:* ________________

(b) $A = 5.5$ cm, $\angle a = 60°$, $\angle b = 30°$. *Answer:* ________________

(c) What is special about this triangle? *Answer:* ________________

Example 7-7: Simpler Solution to Example 3-3

Equipped with these powerful formulas for solving triangles, let us redo a problem we encountered in Chapter 3. We will use the law of sines to find the height of an object. Figure 7.9 shows the measurements. We want to find h.

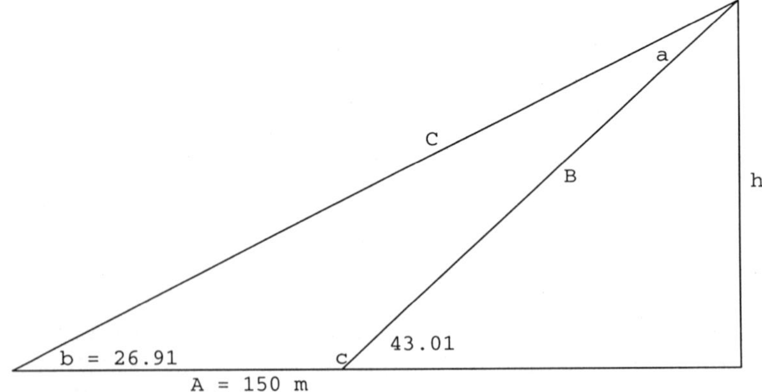

Figure 7.9 Finding the Altitude Using the Law of Sines

You need to know the angles a, b, and c to apply the sine law. Angle $a = 43.01° – 26.91° = 16.1°$. Why is this so? Angle $c = 180° – 43.01° = 136.99°$.

$$\frac{A}{\sin(a)} = \frac{B}{\sin(b)}$$

$$\frac{150}{0.2773} = \frac{B}{0.4526}$$

$$B = 150 \cdot \frac{0.4526}{0.2773} = 244.8$$

$$h = B\sin(43.01°) = 167.0$$

We have found the altitude by the application of the sine law. You should remember that we were able to solve this problem by applying the definitions of the trigonometric functions in Chapter 3. It may take more steps, but you can usually solve a problem in a straightforward manner by applying the fundamental definitions and theorems.

Example 7-8: The Angle between Two Lines

Find the cosine of the angle between the two lines, **A** and **B**, represented as vectors. The components of the vectors are: **A** = [5, 2] and **B** = [6, 3.5].

Solution. The steps for the solution using Maple are the same as in Example 7-5. Here, we will present a solution that directly follows from Equation 7-7: **A**·**B** = 30 + 7 = 37. To find the dot product when you know the components, multiply the x components together and add them to the product of the y components. The magnitudes of **A** and **B** are found using the familiar Pythagorean formula once again: $A = 5.385$ and $B = 6.946$. Thus, $\cos(\theta) = 37/(5.385 \times 6.946) = 0.9892$. Check this result using Maple.

```
> with (linalg);

> A := vector([5, 2]): B := vector([6, 3.5]): cos(theta) = evalf(dotprod(A, B)/(norm(A,
2)*norm(B, 2)));

                    cos(θ) = .9891315765
```

Your Turn: Knowing the cosine of the angle, you can find the angle using the arccos function (Chapter 9). Find the angle this way and measure it in Figure 7.10.

(a) What is the angle as measured in Figure 7.10? *Answer:* ____________________

(b) What is the calculated angle, in degrees? *Answer:* ____________________

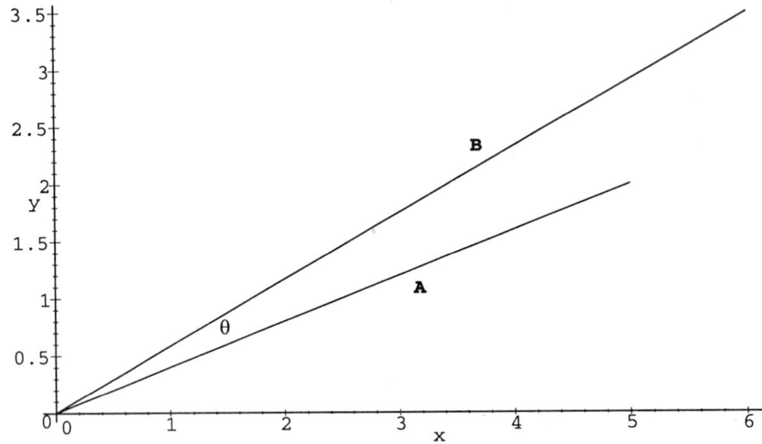

Figure 7.10 The Angle between Two Vectors

Paper and Pencil Exercises

PP7–1

John is lost in an unfamiliar city, where all the blocks are the same size. Trying to find his hotel, he walks 4 blocks north, then turns east and walks 3 more blocks. Still not finding the hotel, he walks 3 blocks south and then 4 blocks west. Beginning to despair, he walks 3 blocks south and one block east, where he finds the hotel.

(a) Sketch his path as he looks for the hotel (he begins at B and ends at E).

Sketch:

(b) Describe the vector BE, giving its magnitude and direction:

Answer: ______________________________

PP7–2

Find the x and y components of the following vectors:

(a) $AB = 10$ at 30° *Answer:* ____________

(b) $CD = 42$ at 60° *Answer:* ____________

(c) $EF = 75$ at 90° *Answer:* ____________

(d) $GH = 163$ at 180° *Answer:* ____________

(e) $IJ = 5\sqrt{2}$ at 225° *Answer:* ____________

PP7–3

Given the x and y components, determine the magnitude and direction of the vector:

(a) $AB_x = 30$, $AB_y = 40$ *Answer:* ____________

(b) $CD_x = 5$, $CD_y = 37$ *Answer:* ____________

(c) $EF_x = -3$, $EF_y = -4$ *Answer:* ____________

(d) $GH_x = -10$, $GH_y = 0$ *Answer:* ____________

(e) $IJ_x = 0$, $IJ_y = -43$ *Answer:* ____________

PP7–4

Add the following vectors. State your answer as a magnitude and a direction:

(a) 10 at 30° + 15 at 90° *Answer:* ____________

(b) $\sqrt{5}$ at 63° 43′ + $\sqrt{5}$ at 26° 56′ *Answer:* ____________

(c) 5 at –60° + 5 at 60° *Answer:* ____________

(d) 2 at 45° + 2 at 90° + 2 at 135° + 2 at 180° + 2 at 225° + 2 at 270° + 2 at 315° + 2 at 360°

Answer: ____________

(e) $\sqrt{2}$ at 45° + $\sqrt{2}$ at 135°+2 at 90° *Answer:* ____________

PP7–5

Evaluate the following:

(a) $\mathbf{A} \cdot \mathbf{B}$ if $\mathbf{A} = 10\mathbf{i} + 13\mathbf{j}$ and $\mathbf{B} = 4\mathbf{i} - 3\,\mathbf{j}$ *Answer:* ____________

(b) **A** · **B** if **A** = 10**i** and **B** = 13 **j** *Answer:* ______________

(c) **A** · **B** if **A** = 10 at 45° and **B** = 10 at 45° *Answer:* ______________

(d) **A** · **B** if **A** = 10 at 0° and **B** = 10 at 90° *Answer:* ______________

(e) **A** · **B** if **A** = $\sqrt{3}$ at 30° and **B** = 2/3 at 0° *Answer:* ______________

PP7–6

True or False?

(a) If **C** = **A** + **B**, is $Cx = Ax + Bx$? *Answer:* ______________

(b) Is | **i** + **j** | = $\sqrt{2}$? *Answer:* ______________

(c) Is | **i**–**j** | = $\sqrt{2}$? *Answer:* ______________

(d) Is | **i**· **j** | = 1? *Answer:* ______________

(e) Is | (**i** + **j**) · (**i**–**j**) | = 2? *Answer:* ______________

PP7–7

If **A** = 3**i** + 4**j**, and **B** = 5**i** + 12**j**, find the cosine of the angle between **A** and **B**: *Answer:* ______________

PP7–8

If **A** = 5**i** and **B** = 3**i**, find **A** × **B**: *Answer:* ______________

PP7–9

If **A** = 5**i** and **B** = 3**j**, find **A** × **B**: *Answer:* ______________

Maple Lab

ML7–1

Given two sides of a triangle as **A** = 14.7**i** and **B** = 12.3 at 57°:

(a) Find the third side, **C**. Hint: load the *linalg* package and issue these commands:

> **A := vector([14.7, 0]);** (Define *A* as a vector)

> **B := vector([12.3 * Cos(57), 12.3 * Sin(57)]);** (Note Sin and Cos, the *degree* functions defined in Prelude to Maple, page 21)

> **C := evalm(A + B);** (Add the vectors)

State why this is a solution to finding the third side of the triangle *ABC*:

Answer: ______________________________

(b) Find side *C* using the cosine law:

> **sqrt(14.7^2 + 12.3^2 + 2 * 14.7 * 12.3 * Cos(57);** (Apply the cosine law directly)

Answer: ______________________________

Note that this last method does not give the direction of vector **C** without further work. If *C* is viewed as just the side of a triangle, it lacks a direction. *C* has only a magnitude, **C** has magnitude and direction.

ML7–2

Given the triangle in Figure 7.11, find the third side, *C*, using the cosine law.

(a) assign directions to **A**, **B**, and **C** as shown in the diagram. Express **A** and **B** as vectors:

Answer: A = ______________________________

Answer: B = ______________________________

(b) What is the angle between **A** and **B**? *Answer:* ____________

(c) Write the Maple command to evaluate *C*: *Answer:* ____________

(d) What is the length of side *C*? *Answer:* ____________

(e) Solve for *C* by vector addition. Define *A* and *B* as

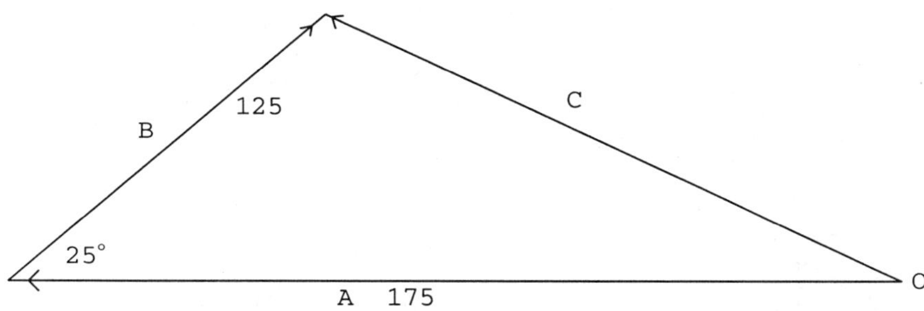

Figure 7.11 Diagram for Problem ML7–2

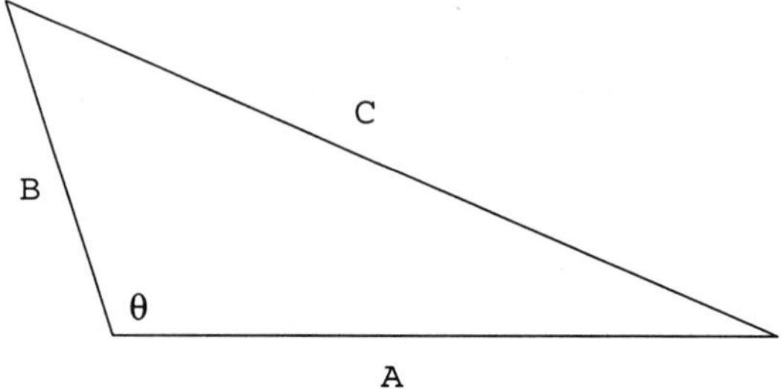

Figure 7.12 Diagram for Problem ML7–3(a)

> **A := vector([Ax, Ay]);** (Put in the proper values for *Ax*, *Ay*)

> **B := vector([Bx, By]);** (Put in the proper values for *Bx*, *By*)

> **C = A + B;** *Answer:* ____________________

ML7–3

Using the method of problem ML7–2, solve the triangles in Figures 7.12 and 7.13 for their third sides:

(a) A = 55, B = 32, θ = 105° *Answer:* ____________________

(b) A = 15, B = 25, θ = 80° *Answer:* ____________________

ML7–4

Given the triangle in Figure 7.13:

Solve for *c*, *a*, and *C*. Assume the other quantities are known.

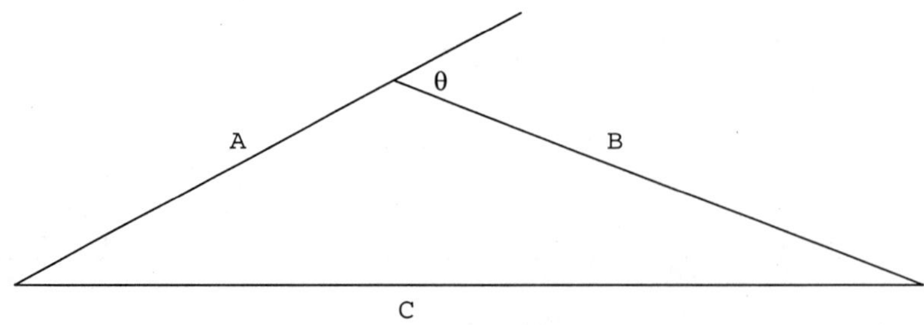

Figure 7.13 Diagram for Problem ML7–3(b)

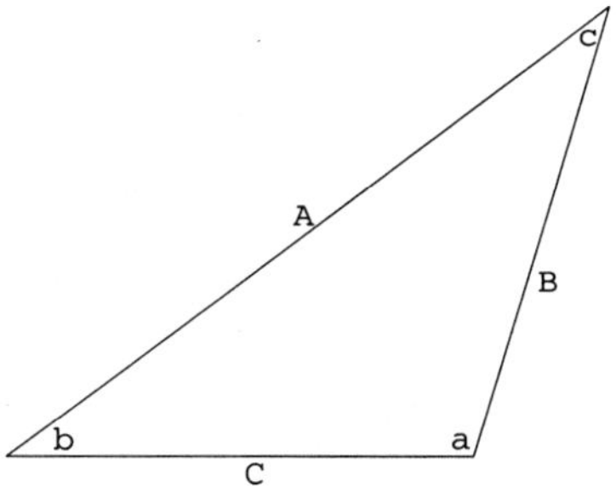

Figure 7.14 Diagram for Problem ML7–4

(a) Use the sine law to calculate c:

> **sr2 :=C/Sin(c) = B/Sin(b);** (Note the use of Sin function. We are able to use degrees for b and c.)

$$sr2 := \frac{C}{\sin(.01745329252c)} = \frac{B}{\sin(.01745329252b)}$$

> **e74 := subs(B=20, C=18, b=37, sr2);**

$$e74 := 18\frac{1}{\sin(.01745329252c)} = 20\frac{1}{\sin(.6457718232)}$$

> **solve(e74, {c});**

$$\{c = 32.79490907\}$$

(b) Once angle c is known, you can solve for angle a:

> **a := 180-32.79-37;**

$$a := 110.21$$

(c) Use the cosine law to solve for side A:

> **A := sqrt(B^2 + C^2 + 2*B*C*Cos(a));**

(d) Solve for A by substituting values for B, C, and a:

> ; (Write the Maple commands here) ____________________

> ;____________________ > ; ____________________

Answers: c = ____________________

a = ____________________

A = ____________________

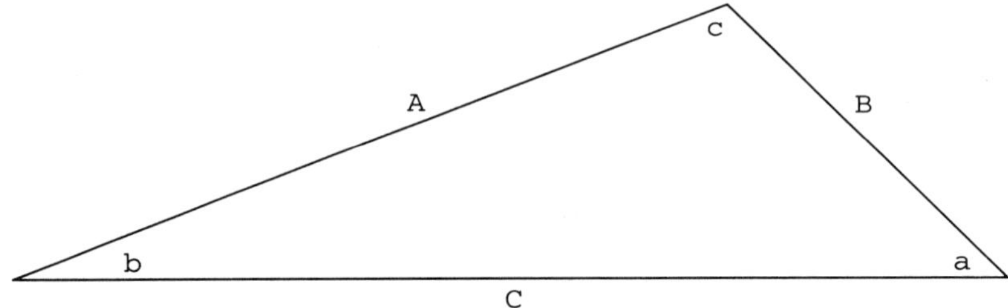

Figure 7.15 Diagram for Problem ML7–5(a)

ML7–5

Use the method of Problem ML7–4 to solve the triangles in Figures 7.15 through 7.17:

(a) See Figure 7.15. A = 8, b = 4, a = 45°

> ; (Write the Maple commands here) ____________________

> ;____________________ > ;____________________

Answers: b = ____________________

c = ____________________

C = ____________________

(b) See Figure 7.16. A = 5.8, B = 5.8, b = 15°

> ; (Write the Maple commands here) ____________________

> ;____________________ > ;____________________

Answers: a =

c = ____________________

C = ____________________

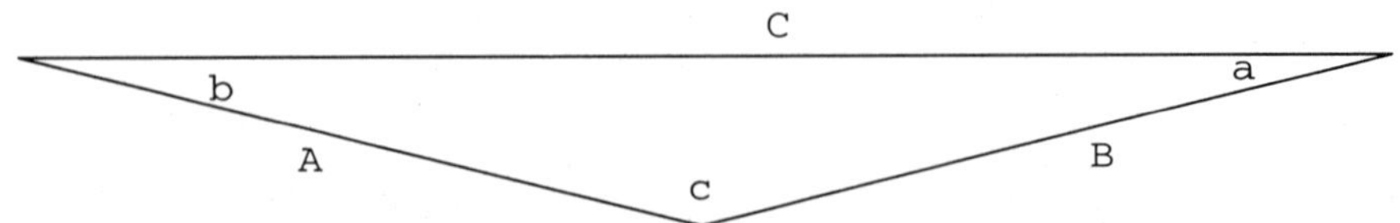

Figure 7.16 Diagram for Problem ML7–5(b)

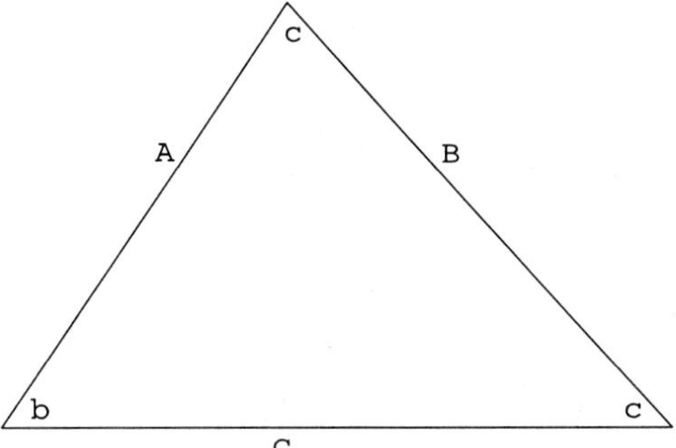

Figure 7.17 Diagram for Problem ML7–5(c)

(c) See Figure 7.17. B = 53.5, A = 48.2, b = 57°

> ; (Write the Maple commands here) ______________________

> ;______________________ > ; ______________________

Answers: a =

c = ______________________

C = ______________________

Trigonometric Functions of Two Angles

Objectives for This Chapter

1. Derive formulas for the sum and difference of sines and cosines using Maple
2. Investigate the double-angle formulas using Maple

Maple Commands Used in This Chapter

evalc	Evaluate a complex number.
expand	Expand a trigonometric expression.
plot(expr, x=0..2, numpoints=500)	Plot a curve using the *numpoints* option.
simplify(expr)	Attempt a simplification.
subs	Substitute values into an expression or equation.

Introduction

Given two angles, there is no difficulty in evaluating the trigonometric functions for the sum of these angles. If the given angles were 30° and 15°, and you wanted the tangent of the sum, you would add the angles and find the tangent. The angles 30° and 15° sum to 45°, and tan(45°) = 1. Similarly, finding the difference of two given angles poses no problem. If you need to determine the sine of the difference between 30° and 15°, subtract the two angles and use Maple to look up the sine function. Since 30° – 15° = 15°, sin(15°) = $\frac{1}{4}\sqrt{6}\left(1-\frac{1}{3}\sqrt{3}\right)$.

The situation changes if the angles are indeterminates. The aim of this chapter is to develop some useful formulas for the sum, difference, and product of two angles. These formulas are useful because they provide a concise way of expressing particular results.

Trigonometric Formulas for Two Angles

Method One:

The most straightforward approach to finding formulas for the sum and difference of two angles is to draw a diagram containing the two angles α and β, and to try and find some relationships between them by drawing construction lines. In Figure 8.1, triangle *BON* contains angle α, and triangle (Δ) *AOB* contains $\angle\beta$. Construct ΔBON first, then draw line $BA \perp OB$. Extend the ray *OA* until it intersects *BA*, forming ΔAOB, which contains angle β at the vertex. This makes ΔOBA a right-angled triangle. The construction lines *AM*, *BN* and *BD* are drawn to form more triangles. Line *BD* is perpendicular to *AM*.

The analysis proceeds by finding relationships among the triangles. The essential feature of this derivation, or "trick," if you like, is to construct line *AB* perpendicular to line *OB*. Who would think of such a thing? Only someone who has thought long and hard about the problem! But once you discover a "trick" like this, you can benefit from it. The approach we have taken in this book is to define angles using the unit circle, in which case the

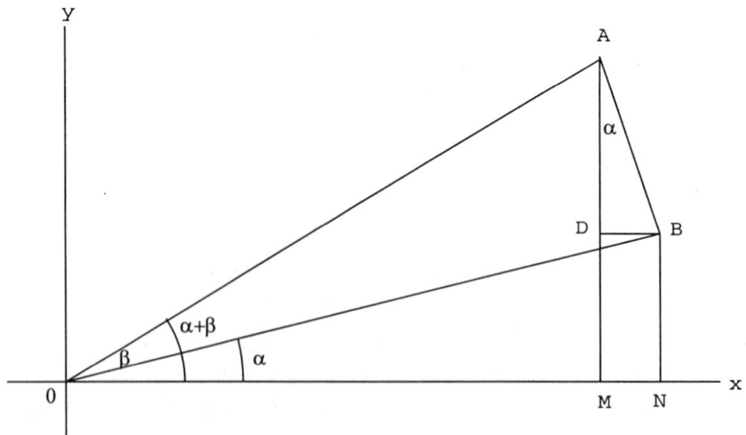

Figure 8.1 Diagram for the Sum and Difference of Two Angles

hypotenuse is always the same length. This is often convenient, but it is not a necessity. The derivation here relies on choosing *OA* to be larger than *OB* by having it meet line *AB*.

$$\sin(\alpha+\beta)=\frac{AM}{OA}=\frac{AD+DM}{OA}=\frac{AD+BN}{OA} \tag{8-1}$$

$$\frac{AD+BN}{OA}=\frac{AD}{OA}+\frac{BN}{OA}=\frac{AD}{AB}\frac{AB}{OA}+\frac{BN}{OB}\frac{OB}{OA} \tag{8-2}$$

$$=\sin(\alpha)\cos(\beta)+\cos(\alpha)\sin(\beta) \tag{8-3}$$

$$\cos(\alpha+\beta)=\frac{OM}{OA}=\frac{ON-MN}{OA}=\frac{ON-BD}{OA} \tag{8-4}$$

$$ON-\frac{BD}{OA}=\frac{ON}{OA}-\frac{BD}{OA}=\frac{ON}{OB}\frac{OB}{OA}-\frac{BD}{AB}\frac{AB}{OA} \tag{8-5}$$

$$=\cos(\alpha)\cos(\beta)-\sin(\alpha)\sin(\beta) \tag{8-6}$$

To derive the cosine formula, we used the fact that $\angle BAD=\alpha$. The proof is as follows: $\angle OBD=\alpha$, since BD is parallel to ON. Since we made $\angle OBA$ a right angle. $\angle DBA=90^\circ-\alpha$; therefore $\angle BAD=\alpha$.

Method Two

This time we will draw the two angles in the unit circle. The approach is based on analytic geometry. The endpoint of the terminal side for angle α is *A*. The coordinates of *A* are $(\cos(\alpha), \sin(\alpha))$, since the terminal side is a radius of the unit circle. Likewise, the point *B* has coordinates $(\cos(\beta), \sin(\beta))$, since it, too, is the endpoint of a radius of the unit circle. There are two "tricks" this time. One is to focus on the length of the line segment *AB*. The other is to realize that the triangle *OAB* in Figure 8.2(a) can be rotated so that *OB* lies along the *x* axis, as shown in Figure 8.2(b).

Under such a rigid rotation, the lengths of the triangle's sides stay the same. Let us write the length of *AB* in both cases, using Maple. Applying the Pythagorean distance formula, we have, from Figure 8.2(a):

$$AB=\sqrt{(\cos(\alpha)-\cos(\beta))^2+(\sin(\alpha)-\sin(\beta))^2} \tag{8-7}$$

> (cos(alpha)-cos(beta))^2 + (sin(alpha)-sin(beta))^2;

$$(\cos(\alpha)-\cos(\beta))^2+(\sin(\alpha)-\sin(\beta))^2$$

We do not need to take the square root. The square of the length, rather than the length itself, works better here. First, we will expand this expression, which is AB^2, and simplify.

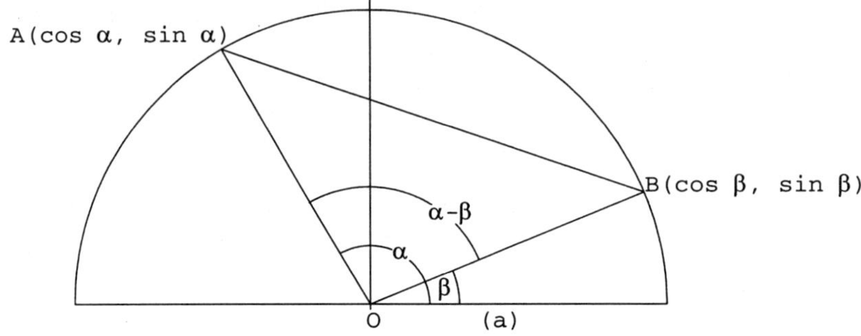

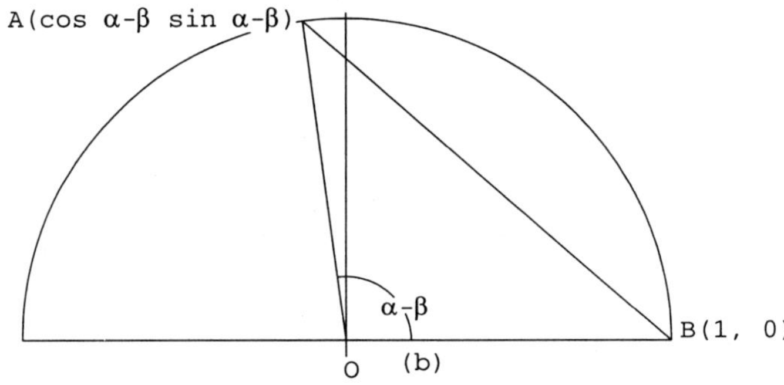

Figure 8.2 Finding cos(α–β) by the Method of Rotation of the Diagram

```
> expand( (cos(alpha)-cos(beta) )^2 + ( sin(alpha)-sin(beta) )^2 );
```

$$\cos(\alpha)^2 - 2\cos(\alpha)\cos(\beta) + \cos(\beta)^2 + \sin(\alpha)^2 - 2\sin(\alpha)\sin(\beta) + \sin(\beta)^2$$

```
> simplify(expand( (cos(alpha)-cos(beta) )^2 + ( sin(alpha)-sin(beta) )^2 ));
```

$$-2\cos(\alpha)\cos(\beta) + 2 - 2\sin(\alpha)\sin(\beta)$$

Here, Maple has used the fact that $\sin(a)^2 + \cos(a)^2 = 1$. We could have done all of this in one line of Maple input by wrapping *expand*, and then *simplify*, around our first expression. The steps can be done in succession by executing the command, observing the result, and then editing the command to produce the next step. In this way, you check your work as you go along. If you prefer to use three separate lines, as we have done here, you will still use the technique of copy and paste. Make sure you get the first input line right by checking Maple's formatted output. Copy the command and paste it on a new input line; then edit it by adding the *expand* command. Copy that input, paste it on the next Maple line, and edit it by adding the *simplify* command. Now you have the desired result for the first part of this derivation.

Now we will apply the same distance formula to the rotated triangle. The distance triangle for *AB*, as shown in Figure 8.2(b), is:

$$AB = \sqrt{(\cos(\alpha - \beta) - 1)^2 + (\sin(\alpha - \beta) - 0)^2} \tag{8-8}$$

As before, we will work with the square of the length, AB^2. This time, we will perform the whole command at once.

> simplify(expand((cos(alpha-beta) - 1)^2 + (sin(alpha - beta) - 0)^2));

$$-2\cos(\alpha)\cos(\beta) + 2 - 2\sin(\alpha)\sin(\beta)$$

This proves that the squares of the lengths are exactly the same, which confirms our typing of the expressions, but it does not give us the identity we are looking for. We want an expression involving the difference of the angles. At this point, why not do the *expand* step and see what happens. Maple expands too far for our purposes, so we employ yet another trick. We call the difference of the two angles γ.

> e1 := subs(alpha - beta = gamma, (cos(alpha-beta) - 1)^2 + (sin(alpha - beta) - 0)^2) ;

$$e1 := (\cos(\gamma) - 1)^2 + \sin(\gamma)^2$$

> expand(e1);

$$\cos(\gamma)^2 = 2\cos(\gamma) + 1 + \sin(\gamma)^2$$

Notice that the sum $\cos(\gamma)^2 + \sin(\gamma)^2$ appears once more. Do you see why you need to memorize this identity? We can remove it by simplifying the expression.

> e1a := simplify(expand(e1));

$$e1a := -2\cos(\gamma) + 2$$

Now we back-substitute for γ.

> subs(gamma = alpha - beta, e1a);

$$-2\cos(\alpha - \beta) + 2$$

Now we have two different expressions for the same length squared, AB^2. Equating them, we find:

$$-2\cos(\alpha - \beta) + 2 = -2\cos(\alpha)\cos(\beta) + 2 - 2\sin(\alpha)\sin(\beta) \tag{8-9}$$

Multiply each side by –1, then add 2, and divide by 2 to obtain:[1]

$$\cos(\alpha - \beta) = \cos(\alpha)\cos(\beta) + \sin(\alpha)\sin(\beta) \qquad (8\text{-}10)$$

Method Three

This method relies on a famous result from complex number theory.[2] Applying the rules of exponentials to $e^{i(\alpha+\beta)}$, we now have the relation:

$$e^{i(\alpha+\beta)} = e^{i\alpha} e^{i\beta}. \qquad (8\text{-}11)$$

The great result from complex analysis is:

$$e^{i\theta} = \cos(\theta) + i\sin(\theta) \qquad (8\text{-}12)$$

where i is the imaginary unit ($i^2 = -1$). One of the ways to ask Maple to work with expressions containing the imaginary unit is the command *evalc*, which evaluates a complex expression to the form $a+ib$. In Maple, the imaginary unit is written as I instead of i.

> exp(I* (alpha + beta)) = evalc(exp(I* (alpha + beta)));

$$e^{(I(\alpha+\beta))} = \cos(\alpha+\beta) + I\sin(\alpha+\beta) \qquad (8\text{-}13)$$

However:

> c1 := exp(I* (alpha + beta)) = expand(exp(I* (alpha + beta)));

$$c1 := e^{(I(\alpha+\beta))} = e^{(I\alpha)} e^{(I\beta)} \qquad (8\text{-}14)$$

We can cause Maple to write this identity in trigonometric form through the use of *evalc*:

> evalc(c1);

$$\cos(\alpha+\beta) + I\sin(\alpha+\beta) = \cos(\alpha) + \cos(\beta) - \sin(\alpha)\sin(\beta) + I(\sin(\alpha)\cos(\beta) + \cos(\alpha)\sin(\beta)) \qquad (8\text{-}15)$$

Complex numbers are equal if, and only if, their real parts are equal and their imaginary parts are equal as well. We can equate the real and imaginary parts separately to get the following equations:

$$\cos(\alpha + \beta) = \cos(\alpha)\cos(\beta) - \sin(\alpha)\sin(\beta) \qquad (8\text{-}6a)$$

---

1. All this could have been done in Maple. If the expression on the right is called *e1b* and the expression on the right is called *e1c*, executing the command:

>(-e1b+2)/2=(-e1c+2)/2;

yields the stated result. Try it!

2. See the companion book, *Maple for Algebra*, Chapter 8, for a discussion of complex numbers.

$$\sin(\alpha + \beta) = \sin(\alpha)\cos(\beta) + \cos(\alpha)\sin(\beta) \tag{8-3a}$$

This is perhaps the easiest derivation to remember. If you know how to multiply complex numbers and the trigonometric form of a complex exponential, you can quickly derive the formulas for cos(α+β) and sin(α+β), as shown in Equations 8-6a and 8-3a.

Method Four

After you try this method, you may wonder why we bothered with the others! If you want to expand cos(α+β) and sin(α+β), use the *expand* command straightaway:

> **cos(alpha + beta) = expand(cos(alpha + beta));**

$$\cos(\alpha + \beta) = \cos(\alpha)\cos(\beta) - \sin(\alpha)\sin(\beta) \tag{8-6b}$$

> **sin(alpha + beta) = expand(sin(alpha + beta));**

$$\sin(\alpha + \beta) = \sin(\alpha)\cos(\beta) + \cos(\alpha)\sin(\beta) \tag{8-3b}$$

Summary of the Methods

We have derived some sum and difference formulas. We used a geometric approach in method one and an analytic geometry approach in method two. Method three relied on the theory of complex numbers, and method four showed that Maple can expand sums of sines and cosines. Why did we show these different methods? You can learn something from each one. Method one is the straightforward approach. You would probably draw a diagram of the problem if you were asked to try and derive a formula for cos(α+β) and sin(α+β). (It probably would look like Figure 8.1.) It might take you a while to realize that the key to the problem (using this approach) is to make *AB* perpendicular to *OB*. When you do, the solution falls into place. Maple is no help in method one.

Method two appeals to those who think dynamically. If you like moving figures around in your imagination, the method of Figure 8.2 will appeal to you. There you find that you must take an algebraic approach. One of the important simplifications is to work with the square of the length *AB* rather than the length *AB* itself. Maple is a help here, because it reliably expands the Pythagorean formula for the length between two points. It helps you keep track of the algebra so that you can apply your thinking skills to the logic behind the derivation: the fact that you can get two formulas, Equations 8-7 and 8-8, for the same length *AB*. From that, the expansion of $\cos(a - b)$ follows.

Method three is for those who have studied the algebra of complex numbers. It is almost like using a crowbar to remove a thumbtack, but if you do know a powerful approach, why not use it?

Method four is fine when (a) you have studied the first two methods to see the derivation from first principles, so that you understand what's involved, and (b) you have Maple handy. Maple is wonderful at remembering formulas. Let it help you to "get the signs right," but realize that you have to supply the understanding!

Sum, Difference, and Product Formulas

We will collect the standard trigonometric formulas in this section for easy reference. In addition, we will show you how to get Maple to tell you these results.

Sum and Difference Formulas for Sine, Cosine, and Tangent

> sin(alpha + beta) = expand(sin(alpha + beta));

$$\sin(\alpha + \beta) = \sin(\alpha)\cos(\beta) + \cos(\alpha)\sin(\beta) \qquad (8\text{-}16)$$

> sin(alpha - beta) = expand(sin(alpha - beta));

$$-\sin(-\alpha + \beta) = \sin(\alpha)\cos(\beta) - \cos(\alpha)\sin(\beta) \qquad (8\text{-}17)$$

> cos(alpha + beta) = expand(cos(alpha + beta));

$$\cos(\alpha + \beta) = \cos(\alpha)\cos(\beta) - \sin(\alpha)\sin(\beta) \qquad (8\text{-}18)$$

> cos(alpha - beta) = expand(cos(alpha - beta));

$$\cos(-\alpha + \beta) = \cos(\alpha)\cos(\beta) + \sin(\alpha)\sin(\beta) \qquad (8\text{-}19)$$

> tan(alpha + beta) = expand(tan(alpha + beta));

$$\tan(\alpha + \beta) = \frac{\sin(\alpha)\cos(\beta) + \cos(\alpha)\sin(\beta)}{\cos(\alpha)\cos(\beta) - \sin(\alpha)\sin(\beta)} \qquad (8\text{-}20)$$

> tan(alpha - beta) = expand(tan(alpha - beta));

$$-\tan(-\alpha + \beta) = -\frac{-\sin(\alpha)\cos(\beta) + \cos(\alpha)\sin(\beta)}{\cos(\alpha)\cos(\beta) + \sin(\alpha)\sin(\beta)} \qquad (8\text{-}21)$$

Double-Angle Formulas

> sin(2*alpha) = expand(sin(2*alpha));

$$\sin(2\,\alpha) = 2\cos(\alpha)\sin(\alpha) \qquad (8\text{-}22)$$

> cos(2*alpha) = expand(cos(2*alpha));

$$\cos(2\,\alpha) = 2\cos(\alpha)^2 - 1 \qquad (8\text{-}23)$$

> tan(2*alpha) = expand(tan(2*alpha));

$$\tan(2\ \alpha) = 2\frac{\cos(\alpha)\ \sin(\alpha)}{2\ \cos(\alpha)^2 - 1} \tag{8-24}$$

Paper and Pencil Exercises

PP8–1

Given $\sin(\alpha) = \frac{2}{3}$ and $\cos(\beta) = \frac{3}{4}$, find:

(a) $\cos(\alpha)$ *Answer:* ____________________

(b) $\sin(\beta)$ *Answer:* ____________________

(c) $\sin(\alpha + \beta)$ *Answer:* ____________________

(d) $\cos(\alpha + \beta)$ *Answer:* ____________________

(e) $\tan(\alpha + \beta)$ *Answer:* ____________________

PP8–2

Given $\cos(\alpha) = \frac{20}{29}$ and $\sin(\beta) = \frac{28}{53}$, find:

(a) $\sin(\alpha)$ *Answer:* ____________________

(b) $\cos(\beta)$ *Answer:* ____________________

(c) $\sin(\alpha - \beta)$ *Answer:* ____________________

(d) $\cos(\alpha - \beta)$ *Answer:* ____________________

PP8–3

Express these sums as products using $2\sin(\alpha)\cos(\beta) = \sin(\alpha + \beta) + \sin(\alpha - \beta)$:

(a) $4\sin(2x)\cos(x)$ *Answer:* ____________________

(b) $2\sin(\omega t)\cos(2\omega t)$ *Answer:* ____________________

(c) $2\sin(x)\cos(x)$ *Answer:* ____________________

(d) $6\sin(4x)\cos(3x)$ *Answer:* ____________________

PP8–4

Express these sums as products using $\sin(\alpha)\pm\sin(\beta) = 2\cos\left(\frac{\alpha\mp\beta}{2}\right)\sin\left(\frac{\alpha\pm\beta}{2}\right)$:

(a) sin(2x) + sin(x) *Answer:* ____________________

(b) sin(x)–sin(2x) *Answer:* ____________________

(c) sin(3x)–sin(2x) *Answer:* ____________________

(d) sin(4x)–sin(3x) *Answer:* ____________________

(e) sin(π/4)–sin(π/6) *Answer:* ____________________

PP8–5

Prove the formulas given in the last two problems.

Maple Lab

ML8–1

Write the result of the following Maple commands. Decide which are simplifications.

(a) expand(sin(2 * alpha + beta)); *Answer:* ____________________

(b) simplify(sin(π/4) - sin(π/6)); *Answer:* ____________________

(c) simplify(sin(x) - sin(2*x)); *Answer:* ____________________

(d) simplify(sin(x) + sin(2*x)); *Answer:* ____________________

(e) simplify(6*sin(4*x)* cos(3*x)); *Answer:* ____________________

ML8–2: Proof of the Fundamental Trigonometric Identities

Use the technique of simplifying the difference of the left- and right-hand sides of the proposed identities. If the result is 0, you have shown that the original equation is an identity, at least according to Maple!

Show that $\csc(x) = 1/\sin(x)$ is an identity. This is obvious, because it is simply the definition of the cosecant function, but it illustrates the process nonetheless. Give the equation a name:

> **eq82 := csc(x) = 1/ sin(x);**

$$eq82 := \csc(x) = \frac{1}{\sin(x)}$$

> **simplify(lhs(eq82)-rhs(eq82));**

$$0$$

Maple says that the difference of the two sides of the equation is 0, so they must be equal. Test the following equations. Answer "true" if the equation is an identity, and "false" otherwise. Note that all the problems after the first involve trigonometric functions of two angles.

(a) > **eq82a := sin(x)^2 + cos(x)^2 = 1;**

> ; (Write the Maple command here) ______________________________

Answer: ______________________________

(b)> **eq82b := cos(u + v) = cos(u)*cos(v)-sin(u)*sin(v);**

$$eq82b := \cos(u + v) = \cos(u)\cos(v) - \sin(u)\sin(v)$$

> ; (Write the Maple command here) ______________________________

Answer: ______________________________

(c)

> **eq82c :=2* sin(x)* cos(y) = sin(x + y) + sin(x-y);**

$$eq82c := 2\sin(x)\cos(y) = \sin(x + y) - \sin(-x + y)$$

> ; (Write the Maple command here) ______________________________

Answer: ______________________________

(d)> **eq82d := 2* sin(x)* sin(y) = cos(x-y)-cos(x + y);**

$$eq82d := 2\sin(x)\sin(y) = \cos(-x + y) - \cos(x + y)$$

> ; (Write the Maple command here) ______________________________

Answer: ______________________________

(e)> **eq82e :=2* cos(x)* sin(y) = sin(x + y)-sin(x-y);**

$$eq82e := 2\cos(x)\sin(y) = \sin(x + y) + \sin(-x + y)$$

> ; (Write the Maple command here) ______________________________

Answer: ______________________________

(f)> **eq82f :=2* cos(x)* cos(y) = cos(x + y) + cos(x-y);**

$$eq82f := 2\cos(x)\cos(y) = \cos(x+y) + \cos(-x+y)$$

> ; (Write the Maple command here) ______________________________

Answer: ______________________________

(g)> **eq82g := sin(x) + sin(y) = 2*cos((x-y)/2) *sin((x+y)/2);**

$$eq82g := \sin(x) + \sin(y) = 2\cos\left(-\frac{1}{2}x + \frac{1}{2}y\right)\sin\left(\frac{1}{2}x + \frac{1}{2}y\right)$$

> ; (Write the Maple command here) ______________________________

Answer: ______________________________

(h)

$$eq82h := \sin(x) - \sin(y) = -2\cos\left(\frac{1}{2}x + \frac{1}{2}y\right)\sin\left(-\frac{1}{2}x + \frac{1}{2}y\right)$$

> ; (Write the Maple command here) ______________________________

Answer: ______________________________

(i)> **eq82i := cos(x) + cos(y) = 2*cos((x+y)/2) *cos((x-y)/2);**

$$eq82i := \cos(x) + \cos(y) = 2\cos\left(\frac{1}{2}x + \frac{1}{2}y\right)\cos\left(-\frac{1}{2}x + \frac{1}{2}y\right)$$

> ; (Write the Maple command here) ______________________________

Answer: ______________________________

(j)> **eq82j := cos(x)-cos(y) = 2*cos((x+y)/2) *sin((x-y)/2);**

$$eq82j := \cos(x) - \cos(y) = -2\cos\left(\frac{1}{2}x + \frac{1}{2}y\right)\sin\left(-\frac{1}{2}x + \frac{1}{2}y\right)$$

> ; (Write the Maple command here) ______________________________

Answer: ______________________________

ML8–3

Consider the formulas:

$$\left|\sin\left(\frac{u}{2}\right)\right| = \sqrt{\frac{1-\cos(u)}{2}}$$

$$\cos\left(\frac{u}{2}\right) = \sqrt{\frac{1 + \cos(u)}{2}}$$

$$\tan\left(\frac{u}{2}\right) = \frac{1 - \cos(u)}{\sin(u)} = \frac{\sin(u)}{1 + \cos(u)}$$

(a) Plot the left- and right-hand sides of these equations over the range $-2\pi \leq u \leq 2\pi$. If you see a single curve, this suggests you are dealing with an identity. Do the left- and right-hand sides yield an identical plot for each of these equations?

> ; (Write the Maple plot commands here) ____________________

Answer: ____________________

(b) Attempt a proof. Use Maple to demonstrate whether these equations are identities. *Hint:* square each side, then simplify the difference.

> ; (Write the Maple plot commands here, showing your steps) ____________________

____

Answer: ____________________

ML8–4: Amplitude Modulation

A signal on the AM band of your radio is produced by impressing a relatively slow audio signal on the carrier frequency. (This is sometimes called the radio wave.) The waveform of such a signal is given by the formula:

> eq84 := A * sin(omega[c]*t) * cos(omega[a] * t);

$$eq84 := A\sin(\omega_c t)\cos(\omega_a t)$$

where A is the amplitude of the radio carrier wave, ω_c is the angular frequency of the carrier wave (the station's assigned frequency on the AM band), and ω_a is the angular frequency of the audio signal. In practice, the carrier frequency is about 1,000 times faster than the audio frequency.

We will present an example where the carrier frequency is only 10 times the audio frequency. You will see why this is so when you plot the graph of this signal.

(a) Substitute the values, $A = 10$, $\omega_c = 100$, and $\omega_a = 10$ and name the new formula *eq84a*. What is the period, T, of the audio signal? (The period, $T = 1/f$ and $\omega = 2\pi f$).

> eq84a := subs(A=10, omega[c]=100, omega[a]=10, eq84):

$$eq84a := 10\sin(100t)\cos(10t)$$

Answer: ____________________

(b) How long does it take for three complete cycles of the carrier wave? Use the formulas in part (a).

Answer: ______________________________

(c) Plot the function for three cycles of the audio frequency. What would the plot look like if you used a carrier frequency of $\omega_c = 100$?

> ; (Write the *plot* command here)

Answer: ______________________________

(d) Sketch the plot. Do you see that the amplitude of the carrier wave is modulated (made greater or smaller) by the audio signal?

Answer: ______________________________

Note: You may have to use the *numpoints* =*500* option in the plot command to get a smooth picture.

Explorations

E8-1: Beats

Suppose a wave is given by the equation

$$S = A\cos(\omega_1 t) + A\cos(\omega_2 t)$$

Examine the case where the two angular frequencies, ω_1 and ω_2, are almost, but not quite, the same. Define the *average frequency* as:

$$\omega = \frac{1}{2}(\omega_1 + \omega_2)$$

and a *modulation frequency* of:

$$\omega_m = \frac{1}{2}(\omega_1 - \omega_2)$$

Use the formulas defined in this chapter to show that:

$$S = 2A\cos(\omega_m t)\cos(\omega 5t)$$

Hint: use Equations 8-18 and 8-19.

E8-2

Here is a challenging problem from the theory of mechanical vibrations. Given:

$$A_{n+1} + A_{n-1} = A_n(2 - K\omega^2)$$

$$A_n = A\sin(kna)$$

Express A_{n+1} as $A\sin(k(n + 1)a)$ and A_{n-1} as $A\sin(k(n - 1)a)$. Simplify these expressions slightly and substitute into the first equation. Since A_{n+1} and A_{n-1} are given as the sum and difference of two angles, use the formulas derived in this chapter to find a simple formula for ω^2 in terms of K and $\cos(ka)$.

Hint: the result you are looking for is $\omega^2 = 2K(1 - \cos(ka))$.

CHAPTER 9

Inverse Trigonometric Functions

Objectives for This Chapter

1. Investigate the meaning of the inverse trigonometric functions
2. Use the definition of the inverse trigonometric functions to solve for angles
3. Find the solution of the inverse problem, where the angle contains additional terms

Maple Commands Used in This Chapter

arcsin(y)	The inverse of y = sin(θ). The angle whose sine is y.
arccos(y)	The inverse of y = cos(θ). The angle whose cosine is y.
arctan(y)	The inverse of y = tan(θ). The angle whose tangent is y.
plot	Plot inverse trigonometric functions.

The Need for the Inverse Trigonometric Functions

Given an angle, you can calculate any of the six trigonometric functions: sin, cos, tan, sec, csc, and cot. Moreover, you can evaluate these trigonometric functions for any angle, large or small, positive or negative. Ask yourself how the process might be turned around. If you know the value of the trigonometric function, how can you determine the angle?

It's not a completely straightforward process. Take the specific example of $\sin(x) = 1/2$. What does this equation imply about the angle x? You can visualize the situation by plotting the sine function $y = \sin(x)$ and the line $y = 1/2$ on the same graph (see Figure 9.1).

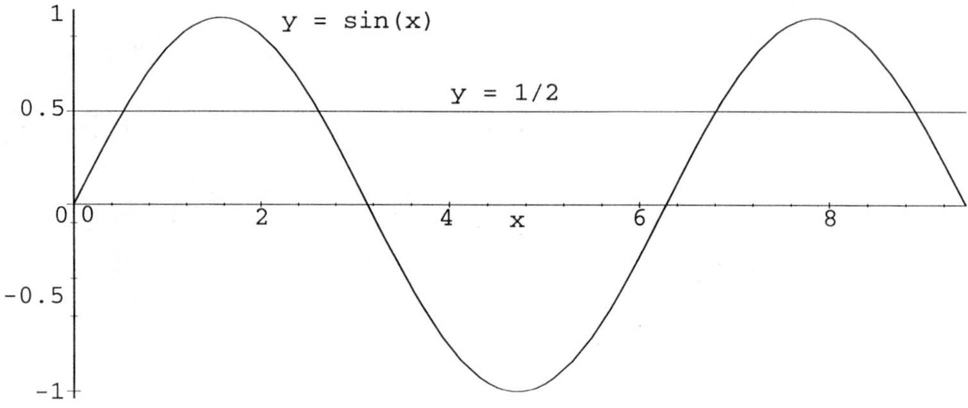

Figure 9.1 The Angles at Which Sin(x) = 1/2

> plot({ sin(x) , 1/2 }, x = 0 .. 3*Pi);

The line intersects the sine curve at four points. Any one of these angles has a sine equal to 1/2. You can continue the sine curve indefinitely in both the positive and negative x direction and the line will continue to intersect the curve. There are an infinite number of solutions to the equation sin(x) = 1/2.

We can arrange the problem of finding the angle, given the trigonometric function, to these steps.

1. Express the problem as an equation.
2. Decide which angles are relevant to the problem. You will need to have additional information about the angle you are looking for because there are an infinite number of possibilities. If the problem doesn't supply this information, you will have to make an assumption and state it as part of your solution.
3. Plot the function and the value on the same graph. Choose the interval you decided upon in step 2. Find the places where the horizontal line intersects the trig function.
4. Read the angles on the x axis that correspond to these intersections. These are the values that satisfy the equation.

Many practical problems involve the determination of the angle given the value of the trigonometric function. We will illustrate the process with an example from surveying. A surveyor is told that a new road is to be built to provide a shortcut from one village to another. The current road travels directly east for 1 mile, where it meets a road going directly north.

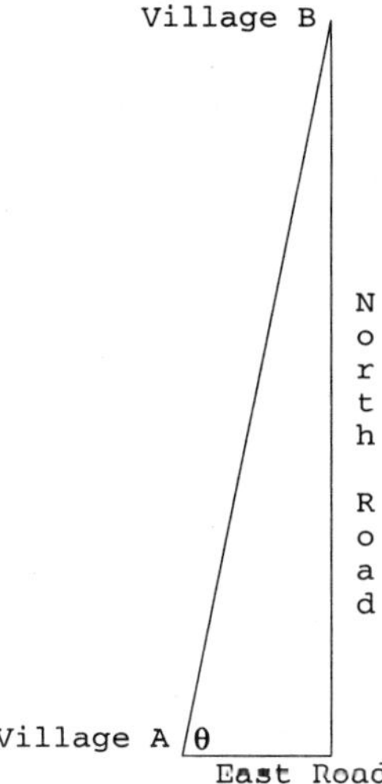

Figure 9.2 The Surveyor's Problem: What Is Angle θ?

The other village is 3 miles up the north road. She is asked to point her transit in the direction of the other village so that the new road can be built along her line of sight. (A transit is a telescope that has a swivel base that measures angles.)

The surveyor is at Village A, and she must point her transit at an angle to the east road. She applies the procedure described above to find the angle (see Figure 9.2).

Step 1: The east road is 1 mile long and the north road is 3 miles long; therefore, the appropriate trigonometric equation is $\tan(\theta) = 3/1 = 3$.

Step 2: The angle θ must be between 0 and 90°. (Angles on a transit are not measured in radians.) Also, the angle must be positive.

Step 3: Plot the trigonometric function and its value. Since the tangent function can grow to be very large near 90°, restrict the y axis to just above 3.

> **plot({ tan(theta), 3 }, theta = 0 .. Pi/2, 0 .. 3.5);**

Step 4: The angle, as read from the graph, is $\theta = 1.25$. We have used Maple's tangent function, so the angle is in radian measure. The angle in degrees is, therefore, $\theta = 1.25(180/\pi) = 71.6°$ (see Figure 9.3).

Finding the angle given the trigonometric value is the inverse to the problem of finding the trig value given the angle. In fact, we have used these inverse functions every time we found the angle, given the lengths of the sides of a triangle, as in the problem above. This "inverse

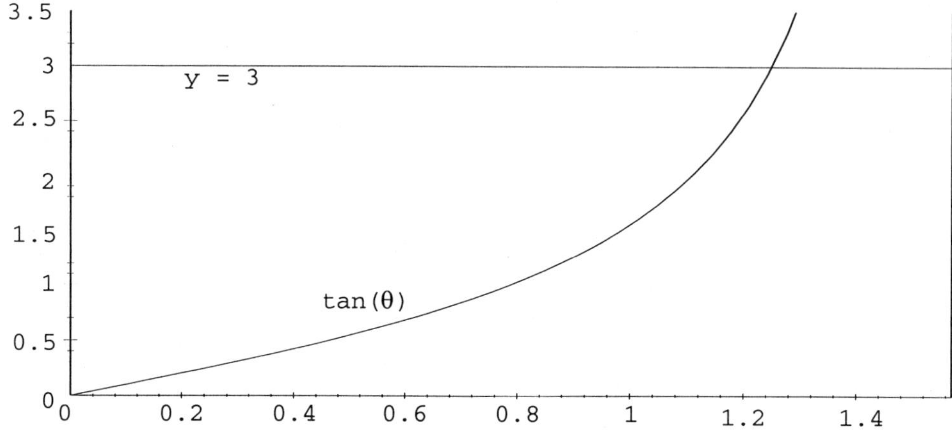

Figure 9.3 Solution to the Surveyor's Problem, tan(θ) = 3

problem" occurs frequently enough that these functions have been given names. They are arcsin, arccos, and arctan. They are defined in terms of the corresponding sin, cos, and tan functions.

$$\theta = \arcsin(y) \text{ corresponds to } y = \sin(\theta) \tag{9-1}$$

$$\theta = \arccos(y) \text{ corresponds to } y = \cos(\theta) \tag{9-2}$$

$$\theta = \arctan(y) \text{ corresponds to } y = \tan(\theta) \tag{9-3}$$

If these inverse functions are truly to be functions, we must restrict the possible values of θ; otherwise they will not be single valued.

The graphs of these functions are obtained by rotating the corresponding sin, cos, and tan graphs by 90° (see Figures 9.4, 9.5, and 9.6).

```
> plot( arcsin(y), y = -Pi/2 .. Pi/2);
> plot( arccos(y), y = -Pi/2 .. Pi/2);
> plot( arctan(y), y = -Pi/2 .. Pi/2);
```

The graph in Figure 9.6 does not tell the whole story about the arctan function. All of these inverse functions will be investigated in detail in the Maple lab for this chapter.

Example 9-1

Solve Example 3-5 using the inverse trigonometric functions.

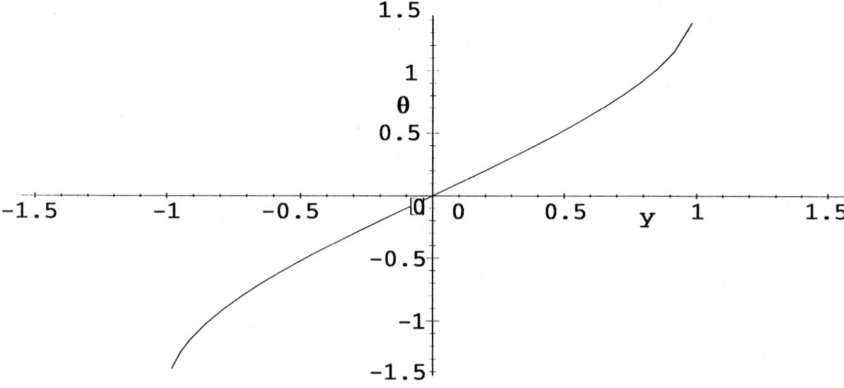

Figure 9.4 Graph of $\theta = \arcsin(y)$

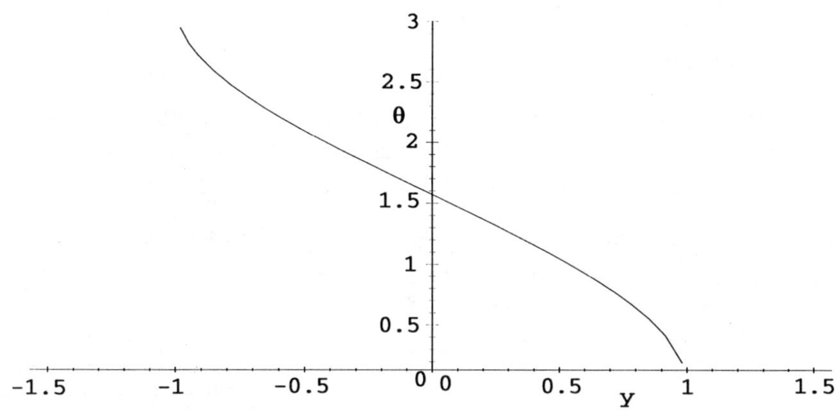

Figure 9.5 Graph of $\theta = \arccos(y)$

Solution. The diagram for this problem is shown in Figure 5.6, where $y = 6.23$ and $x = 14.5$. Since $\tan(\theta) = 6.23/14.5 = 0.430$, $\theta = \arctan(0.430) = 23.7°$. The angle $\alpha = 90° - \theta\ 90 - 23.27 = 66.73°$, and $r = x/\cos(\theta) = 15.8$.

Your Turn. Solve Example 3-5 using the inverse trigonometric functions where $y = 10.22$ and $x = 7.35$.

Answer: ____

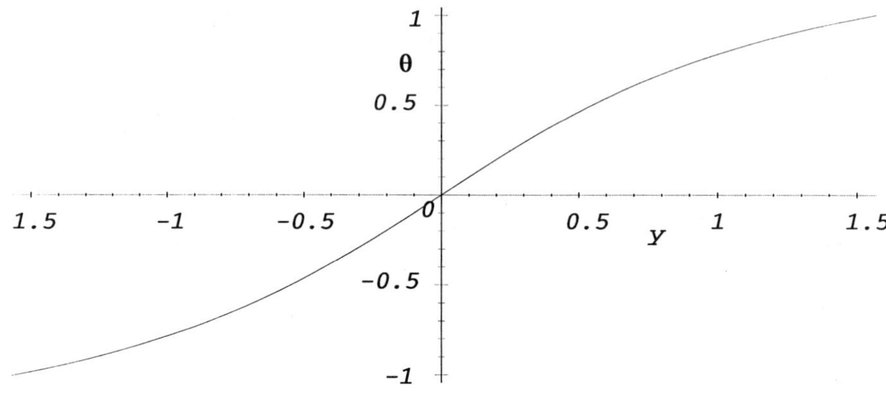

Figure 9.6 Graph of $\theta = \arctan(y)$

Example 9-2

Solve for θ in Example 5-5.

Solution. Review Example 5-5 to see that we derived the equation $\cos(\theta) = 0.7054$. Knowing the value of the cosine function, we find θ by evaluating $\theta = \arccos(0.7054) = 45.14°$. This is the same answer that we found by trial and error in Example 5-5.

Your Turn. Solve Example 5-5 for θ using the inverse trigonometric functions where $a = 55.3$ m, $b = 75.25$ m, and $c = 127$ m. After you have found θ, find h.

Answer: $\theta =$ ______________________________

Answer: $h =$ ______________________________

Finding the Time When a Trigonometric Function Reaches a Specified Value

Consider a problem that you will encounter if you use trigonometric functions to describe oscillations of some kind. They may be mechanical vibrations, or alternating current signals in electronics. The list of possible applications is very large.

Here is a typical example:

An alternating current is given by the formula

$$i(t) = A \sin(\omega t + \alpha)$$

$$i(t) = 110\sin(120\pi t + 1.600)$$

The *amplitude* of the current is 110 milliamperes (mA) and the frequency is 60 hertz (Hz). There is also a *phase* term (α), which in this case is 1.600 radians.

We want to find the *smallest positive* time when the current is 55 mA.

Solution: the first steps pose no problem. We set up the equation as:

```
> eq1 := 55 = 110*sin(120*Pi*t + 1.6);
```

$$eq1 := 55 = 110\sin(120\pi t + 1.6)$$

This equation is equivalent to

```
> eq2 := sin(120*Pi*t + 1.6) = 0.5;
```

$$eq2 := \sin(120\pi t + 1.6) = .5$$

Since we can define an angle θ as $\theta = 120\,\pi t + 1.6 - 120\,\pi t + 5\,\pi/6$, we arrive at the simple equation

```
> sin(theta) = 1/2:
```

The solution for θ is:

```
> theta = arcsin(1/2);
```

$$\theta = \frac{1}{6}\pi$$

Substituting for θ:

```
> eq3 := subs(theta = Pi/6, 120 * Pi * t + 5*Pi /6 = theta);
```

$$eq3 := 120\pi t + \frac{5}{6}\pi = \frac{1}{6}\pi$$

We now can solve for t:

```
> solve( eq3, t);
```

$$\frac{-1}{180}$$

We have found t to be $-1/180$ s. We have *a* solution (one of an infinite number), but the time is negative rather than positive!

How can we use this solution to find all the other solutions? In particular, how do we use this solution to find the one for the smallest positive time? Note what we have found: the angle θ is $\pi/6$, or 30°. This is the smallest positive angle that satisfies the equation $\sin(\theta) = 1/2$. If θ increases, will there be other values for which $\sin(\theta) = 1/2$? Think of the angle getting larger as time goes on. After all, this is what the original equation is saying! At some time after θ was $\pi/6$, it will have made a further full oscillation. We add 2π to $\pi/6$ to get $\theta = \pi/6 + 2\pi$. Check the sine of this new angle using Maple:

> sin(Pi/6 + 2*Pi);

$$\frac{1}{2}$$

Maple immediately reduces the angle to the reference angle, and calculates that the result is still 1/2. But do we need to execute one full oscillation before the sine function repeats the value 1/2? Examine the two angles θ and $\pi-\theta$ in Figure 10.2 (p. 216):

The value of the sine function is y/r. Any height y in Quadrant 1 (Q1) has an equal height in Q2 and, therefore, the same sine. If the smaller angle is θ, the larger angle is $\pi - \theta$. Check the diagram to verify this fact. We see that we have a list of successively larger angles, θ, $\pi - \theta$, $\theta + 2\pi$, $3\pi - \theta$, $\theta + 4\pi$, and so on. All these angles solve the original equation.

If we use $\pi - \theta$ instead of θ in our problem here, we arrive at the equation (remember, we found θ to be $\pi/6$):

> eq4 := 120*Pi*t + 5*Pi/6 = Pi - Pi/6;

$$eq4 := 120\pi t + \frac{5}{6}\pi = \frac{5}{6}\pi$$

> solve(eq4, t);

$$0$$

Thus, the current is 55 mA at $t = 0$. Since 0 is sometimes considered to be a positive number, it is the smallest positive time when the current is 55 mA.

What is the next later time when the current is again 55 mA? Use the next value in the sequence for the angle:

> eq5 := 120*Pi*t + 5*Pi/6 = 2*Pi + Pi/6; solve(eq5, t);

$$eq5 := 120\pi t + \frac{5}{6}\pi = \frac{13}{6}\pi$$

$$\frac{1}{90}$$

Thus, $t = 1/90$ s (or $t = 0.01111$ s or 11.1 ms).

Review: to solve the equation $A \sin(\omega t + \alpha)\ B$ for t:

1. Let $\theta = \omega t + \alpha$. The given equation simplifies to $A \sin(\theta) = B$.

2. Write the equation as $\sin(\theta) = \frac{B}{A}$, $\therefore \theta = \arcsin\left(\frac{B}{A}\right)$ and solve for θ.

3. Realize that if θ is a solution, so is $\pi - \theta$, $\theta + 2\pi$, $3\pi - \theta$, $\theta + 4\pi$, and so on. You can derive a sequence of equations whose right-hand side constantly increases as the sequence continues. As you subtract α from an increasingly larger quantity, you must reach a point where the result is 0 or a positive number. At that point you have the solution for the smallest positive time.

Paper and Pencil Exercises

PP9–1

Assume $-\pi \leq \theta < \pi$. Find θ in radians, given:

(a) $\sin(\theta) = 0.4379$ *Answer:* ____________

(b) $\cos(\theta) = 0.0037$ *Answer:* ____________

(c) $\tan(\theta) = 4.379$ *Answer:* ____________

(d) $\sin(\pi - \theta) = 0.5555$ *Answer:* ____________

(e) $\cos(\pi - \theta) = -0.4349$ *Answer:* ____________

(f) $\tan(\pi - \theta) = -1.732$ *Answer:* ____________

PP9–2

Assume $-180 \leq \theta < 180$. Find θ in degrees, given:

(a) $\sin(\theta) = 0.4866$ *Answer:* ____________

(b) $\cos(\theta) = -0.5$ *Answer:* ____________

(c) $\tan(\theta) = 3.732$ *Answer:* ____________

(d) $\sin(90 - \theta) = 0.9659$ *Answer:* ____________

(e) $\cos(90 - \theta) = 0.1736$ *Answer:* ____________

(f) $\tan(90 - \theta) = 0.1763$ *Answer:* ____________

PP9–3

Find θ, knowing that it is in a given quadrant. Answer in degrees.

(a) Q4 and $\sin(\theta) = -0.139$ *Answer:* ____________

(b) Q3 and $\cos(\theta) = -0.9397$ *Answer:* ____________

(c) Q2 and $\tan(\theta) = -1.428$ *Answer:* ____________

PP9–4

Find θ, knowing that it is in a given quadrant. Answer in radians.

(a) Q3 and $\sin(\theta) = -0.93969$ *Answer:* ____________

(b) Q4 and $\cos(\theta) = 0.01745$ *Answer:* ____________

(c) Q2 and $\tan(\theta) = -2.747$ *Answer:* ____________

Maple Lab

ML9–1

Find θ, given

$$\frac{\sin(\theta) + 0.5}{\cos(\theta)^2 + 1.4} = 0.4153$$

(a) Find one value of θ that solves the equation by Maple's algebraic methods:

Answer: ____________

(b) Verify your solution by plotting and find at least two other graphical solutions:

Answer: ____________

ML9–2

Use the formula $i(t) = I_{max} \sin(2\pi ft)$ and Maple to find:

(a) i, given $I_{max} = 50$ A, $f = 1000$ Hz, and $t = 0.0025$ s (2.5 ms).

(b) t, given $I_{max} = 10$ mA, $f = 10$ kHz, and $i = 1$ mA. Is your solution the smallest positive t that satisfies the given conditions?

Answer: ____________

(c) f, given $i = 2$ mA, $I_{max} = 10$ A, $t = 0.001$ s (1 ms). Is your solution the smallest frequency that satisfies the equation? Note that f must be a positive number.

Answer: ____________

ML9–3

During an earthquake, the displacement of a point on a bridge was given by

$$d = 100\sin(250\pi t + \frac{1}{3}\pi)$$

The maximum displacement was 100 mm (approximately 4 in). (Displacement is the distance of a point from its undisturbed location.)

(a) Find the displacement at $t = 30$ ms. *Answer:* ____________________

(b) Find the smallest positive time for which $d = 45$ mm.

Answer: ____

ML9–4

Electronics engineers know it is possible to express complex wave patterns as sums of sine terms. Consider the following complex wave pattern:

$$v(t) = 10\ \sin(120\pi t) + 3.1\ \sin(240\pi t) + 7.5\ \sin(360\pi t)$$

Find a graphical solution for the smallest positive time when the voltage $v(t)$ is equal to 15 volts (V). Here are the steps you should follow:

1. Name the equation v. *Answer:* ____________________
2. Use the *plot* command: > plot(v, t = -1/180 .. 1/60);.
3. Read the solution from the graph. *Answer:* ____________________

Explorations

Think about how you might try for an algebraic solution to the last question, ML9–4. You know that $\sin(2x) = 2\sin(x)\cos(x)$. As a preliminary to this exploration, see if you can show that

$$\sin(3x) = 2\ \sin(x)\ \cos(x)^2 + \sin(x)(\cos(x)^2 - 1)$$

1. Make the substitution $x = 120\ \pi t$ in $v(t) = 10\ \sin(120\pi t) + 3.1\ \sin(240\pi t) + 7.5\ \sin(360\pi t)$.
2. Express $\sin(240\ \pi t) = \sin(2x)$ in terms of $\sin(x)$ and $\cos(x)$.
3. Express $\sin(360\ \pi t)$ in terms of $\sin(x)$ and $\cos(x)$.

4. Attach the given amplitudes and simplify the resulting equation.
5. Attempt to solve the problem explicitly, or use the *fsolve* command.
6. Having solved for x, find t.
7. Comment on your solution. If all you were after was the value for the smallest positive time, was the analytic solution any better than the graphical solution?

CHAPTER

10

Trigonometric Equations

Objectives for This Chapter

1. Find a single solution to a trigonometric equation
2. Apply rules for generating all solutions to find those that apply to a given domain
3. Apply methods of solution, symbolic, graphical, and numerical to solve trigonometric equations using Maple

Maple Commands Used in This Chapter

arccos(A)	The angle whose cosine is A.
arcsin(A)	The angle whose sine is A.
arctan(A)	The angle whose tangent is A.
cos(theta)	The cosine of angle θ, where θ is in radian measure.
evalf	Convert solutions containing π to approximate decimals.
plot	Use Maple's *plot* function to visualize the solution to a trigonometric equation.
simplify	Use *simplify* on solutions to trigonometric equations to see if any additional simplification in the answer results.
sin(theta)	The sine of angle θ, where θ is in radian measure.
solve	Maple's *solve* command works for most simple trigonometric equations.
tan(theta)	The tangent of angle θ, where θ is in radian measure.

Many Solutions from One

Trigonometric equations usually have infinitely many solutions. This general result is easily seen by considering a simple problem. Suppose you have solved a trigonometric problem and have found that the angle θ you are looking for is given by the equation

$$\sin(\theta) = \frac{3}{5} \tag{10-1}$$

What are the possible values of θ?

You know that the solution involves the inverse function, arcsin(θ), so you can use your calculator:

$$\sin^{-1}\left(\frac{3}{5}\right) = 36.87^\circ \text{ (degree mode)}$$

$$\sin^{-1}\left(\frac{3}{5}\right) = 0.6435 \text{ (radian mode)} \tag{10-2}$$

Your calculator gives you only one value for the inverse sine, no matter whether your calculator is in degree mode or in radian mode.

Maple is even less instructive. If you issue the command:

> **arcsin(3/5);**

$$\arcsin\left(\frac{3}{5}\right)$$

it merely repeats the command without giving you any further information. Of course, it is really telling you that there is no *exact* angle whose sine is 3/5. If you ask for the decimal approximation:

> **evalf(arcsin(3/5));**

$$.6435011088$$

you get the same answer as you did using the calculator.

Are these the only possible answers? Remember how the arcsin function is described in words: *the angle whose sine is 3/5*. Thus, we are looking for the solutions to the equation

$$\sin(\theta) = \frac{3}{5}$$

The graphical solution is found by plotting both sides of the Equation 10-3 over a fairly broad domain for θ, such as θ = –4*Pi .. 4*Pi (see Figure 10.1).

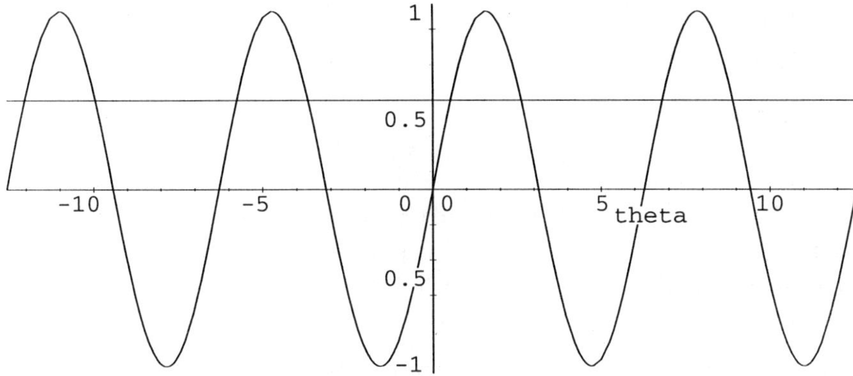

Figure 10.1 Some Solutions of $\text{Sin}(\theta) = \frac{3}{5}$

```
> plot( { sin(theta), 1/2 }, theta = -4*Pi .. 4*Pi);
```

You can see that the horizontal line at 0.6 (or 3/5) crosses the curve for sin(θ) at many places. If the domain for θ were extended, the line would cross more times. The sine function oscillates back and forth, between −1 and +1, infinitely many times as θ is allowed to grow very large in either the positive or the negative direction.

These angles are all related; in fact, there are two relations. The simpler of the two relations comes from the observation that any angle greater than 360° (or 2π radians) is equivalent to an angle in the domain $0 \le \theta < 360°$, or $0 \le \theta < 2\pi$ radians. Thus, you can add or subtract any multiple of 360° (or 2π radians) to any solution for θ in $\sin(\theta) = \frac{3}{5}$ and the result will also be a solution. Try it on your calculator: 360 + 36.87 = 396.87, and sin(396.87) = 0.6, or 3/5. Equivalently, you can test the solution using Maple. We will take the sine of a number of full rotations plus the original angle. Study the following command to see that we are taking the exact sine value of four rotations plus the original angle:

```
> sin(8*Pi + arcsin(3/5));
```

$$\frac{3}{5}$$

The answer is still 3/5. If we took 100 negative rotations, the answer would still be the same.

```
> sin(-200*Pi + arcsin(3/5));
```

$$\frac{3}{5}$$

In the last Maple command, what, really, is the angle whose sine we are taking? Add the two terms and express them as an approximate decimal to obtain:

> evalf(-200*Pi + arcsin(3/5));

```
-627.6750297
```

We are taking the sine of –627.675 radians and finding that it is equal to 3/5, so it is a solution to Equation 10-3. When we express the angle as an approximate decimal, it is not so obvious that the angle is related to the reference angle by subtracting 100 full rotations. Try the command:

> for i from -10 to 10 do sin(-2*i*Pi + arcsin(3/5)) od;

which evaluates the sine of the angle for 21 different cases. It begins the evaluation by adding 10 negative rotations to the reference angle, then –9, and so on, all the way up to 10 positive rotations. In each case, the result is 3/5. Try it and verify the last statement.

There are other solutions as well. These solutions arise from the fact that trigonometric function values repeat themselves in two quadrants of the basic circle. Figure 10.2 shows that there is an angle in Quadrant 2 with the same sine as an angle in Quadrant 1.

Since all angles are measured with reference to the x axis, the angle θ has the same sine as angle $\pi - \theta$. You can use the unit circle to show that $\cos(\theta) = \cos(2\pi - \theta)$ and tan((θ) = tan(π

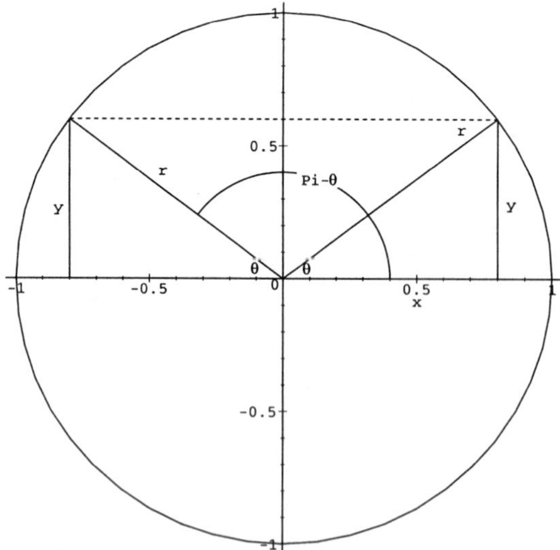

Figure 10.2 Two Triangles with the Same Sine

+ θ). Of course, any number of full revolutions can be added to these new angles and they will still satisfy the original equation.

Example 10-1

Solve the equation $\tan(\theta) = \frac{1}{\sqrt{3}}$ and state all possible angles.

Solution. We will present a Maple V4 assisted solution:

> solve(tan(theta) = 1/sqrt(3), theta);

$$\frac{1}{6}\pi$$

Note that Maple Release 3 does not find the answer. Instead, it simply says that the solution is:

$$\arctan\left(\frac{1}{3}\sqrt{3}\right) \tag{10-4}$$

which is correct, but not very helpful.

Assuming that we have the solution as $\theta = \pi/6$, we form the other solutions:

> x :=tan(Pi/6): x-6*Pi, x-4*Pi, x-2*Pi, x-0*Pi, x+ 2*Pi, x+ 4*Pi, x+ 6*Pi, x+ 8*Pi;

$$s1 := \frac{1}{3}\sqrt{3} - 6\pi,\ \frac{1}{3}\sqrt{3} - 4\pi,\ \frac{1}{3}\sqrt{3} - 2\pi,\ \frac{1}{3}\sqrt{3},\ \frac{1}{3}\sqrt{3} + 2\pi,\ + \frac{1}{3}\sqrt{3} + 4\pi,\ \frac{1}{3}\sqrt{3} + 6\pi,\ \frac{1}{3}\sqrt{3} + 8\pi$$

> s2 := -3*Pi + x, -Pi + x, x, x + Pi, x + 3*Pi, x + 5*Pi, x + 7*Pi;

$$s2 := -3\pi + \frac{1}{3}\sqrt{3},\ -\pi + \frac{1}{3}\sqrt{3},\ \frac{1}{3}\sqrt{3},\ \frac{1}{3}\sqrt{3} + \pi,\ \frac{1}{3}\sqrt{3} + 3\pi,\ \frac{1}{3}\sqrt{3} + 5\pi,\ \frac{1}{3}\sqrt{3} + 7\pi$$

Take any of the solutions in s1 or s2, for instance, take s1[2] and s2[6]. Ask Maple for the tangent:

> tan(s1[2]), tan(s2[6]);

$$\tan\left(\frac{1}{3}\sqrt{3}\right),\ \tan\left(\frac{1}{3}\sqrt{3}\right)$$

Maple reports that these angles have the same tangent as $\tan(\pi/6)$, namely, $\sqrt{3}/3$. As far as the tangent function is concerned, all these angles are the same.

We have investigated the apparently simple problem of finding all the angles corresponding to a known tangent value because most people find the topic confusing. It is simple to look up a trigonometric function given an angle: there is only one answer. The inverse problem is complicated by the fact that there may be many possible answers. Use Equation 10-5 to generate the solutions that lie "in the other quadrant."

$$\sin(n\pi - \theta) = \sin(\theta)$$

$$\cos(n2\pi - \theta) = \cos(\theta)$$

$$\tan(n\pi + \theta) = \tan(\theta) \qquad (10\text{-}5)$$

In Equation 10-5, n can be any *odd* integer

Your Turn: Use Maple to evaluate Equation 10-5 for $n = 7$. Since Maple has basic trigonometric simplification routines built in to its deepest level, you don't need to issue the *simplify* command. What is Maple's response when you type:

> sin(13*Pi - theta); *Answer:* ____________________

> cos(26*Pi - theta); *Answer:* ____________________

> tan(13*Pi + theta); *Answer:* ____________________

Observe that these commands are all examples of Equation 10-5.

Specifying the Domain Limits Solutions

Most problems dealing with unknown angles give you some hint about the domain. This information helps you find a solution beginning with the graphical approach. Let us say that you want a solution to the problem illustrated in Figure 10.3. The diagram shows the path of a ray of light as it travels from the upper right, enters a transparent substance like glass, and continues on through the glass to the lower left. It is a physical fact that the light ray bends when it enters a different substance at any angle other than head-on.

The relation between the angle of incidence, i; the angle of refraction, r; and the *index of refraction*, n, is given by Snell's law:

$$\frac{\sin(i)}{\sin(r)} = n \qquad (10\text{-}6)$$

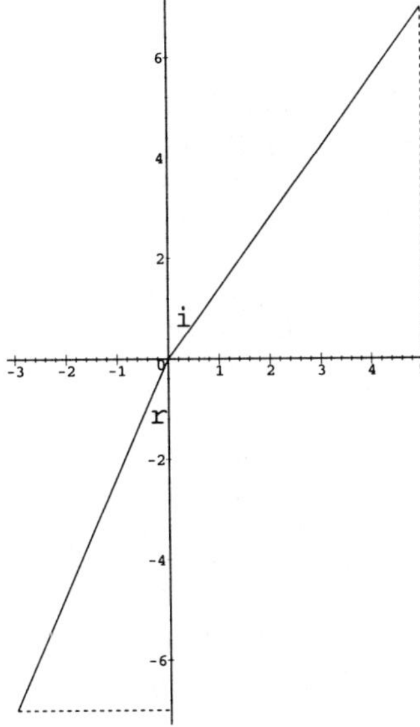

Figure 10.3 The Refraction of a Ray of Light

Example 10-2

Find the index of refraction, n, for the conditions shown in Figure 10.3.

Solution. From the graph, $\tan(i) = 7/5$, $\tan(r) = 3/7$, we see that the angles must be less than 90°. The angles are found using the inverse tangent function, $i = \arctan(7/5)$, $r = \arctan(3/7)$. Since Maple's arctan function returns an angle between $-\pi$ and π,

> i = evalf(180*arctan(7/5)/Pi);

$$i = -54.46232221$$

> r = evalf(180*arctan(3/7)/Pi);

$$r = 23.19859051$$

You can determine n from the Maple command:

> n = evalf(sin(arctan(7/5))/sin(arctan(3/7)));

$$n = 2.065736495$$

Note that in this example, neither angle is a reference angle because both are defined in terms of the y axis rather than the positive x axis. In fact, i is defined with reference to the positive y axis, and r is defined with reference to the negative y axis.

Example 10-3

You are given a material whose index of refraction, n, is 1.4, and you are asked for the angle of refraction, r, if the angle of incidence, i, is 75°.

Solution. Rearrange Equation 10-6 to obtain $\sin(r) = \sin(i)/n$. Using Maple to apply the inverse function:

> r = evalf(180/Pi*arcsin(sin(Pi*75/180)/1.4));

$$r = 43.62591508$$

The angle of refraction is about 43.6 degrees. Note how the angles were converted to radians or degrees when necessary, and note especially how the arcsin function was used to find r. You could have done the problem in three steps, first finding $\sin(i)/n$, then evaluating the arcsine to find r in radians, and finally converting r back to degrees. That is a good way to do the problem. Just realize that you can build up the above Maple input line by using the three steps in turn. Each successive command becomes a "wrapper" for the enclosed commands.

Your Turn: Find angle i if $n = 1.432$ and $r = 10.7°$. *Answer:* ____________________

Example 10-4

An alternating current (AC) voltage source is given by the equation:

$$v(t) = A\sin(\omega t + \alpha) \quad (10\text{-}7)$$

The *amplitude* of the voltage wave is A and is measured in volts. The *angular frequency* is ω, measured in radians per second, and α is the *phase,* measured in radians. A practical problem is to find the smallest *positive* time when the voltage first reaches some value. The data are as follows: $A = 17.5$ V, $\omega = 376.99$ rad/s, and $\alpha = 5.535$ rad. What is the minimum positive time when the voltage, $v(t)$, attains a value of +5 V?

Solution. Always plot the problem if you can. In this case, you can quickly obtain an approximate solution with the command:

> plot({5, 17.5*sin(376.99*t + 5.535) }, t = 0 .. 0.02);

The sine wave rises to +5V at approximately 0.00275 s (2.75 ms) (see Figure 10.4). Now we give the analytical solution, using Maple. The first step is to simplify the angle by making the substitution:

> e3a := theta = 376.99*t + 5.535:

Next, solve for theta:

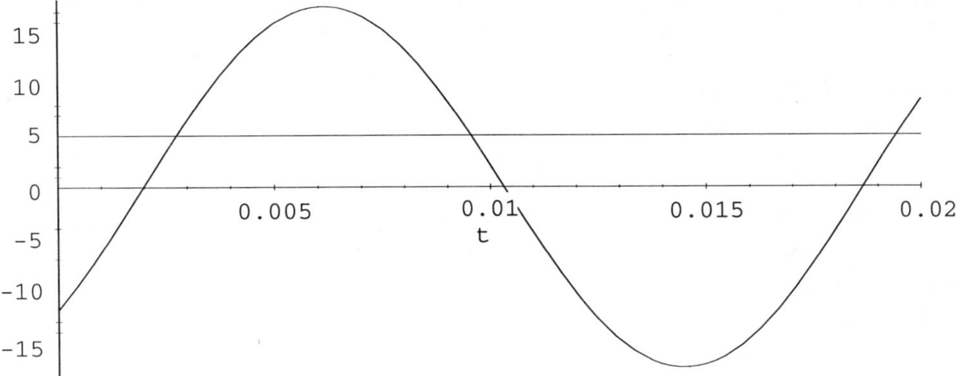

Figure 10.4 Graphical Solution for the Equation 17.5 sin (376.99t + 5.535) = 5

> theta := arcsin(5/17.5);

```
theta := .2897517014
```

Be careful of the next step! We have equation e3a, giving *t* in terms of *theta*, so you might think that all that needs to be done is to solve for *t*:

> solve(e3a, t);

```
-.1391349452e-1
```

The answer is negative! This is not the value of *t* we want. (It is *one* solution, however.) Since we want a later time, we need to use a bigger angle. The next bigger angle is obtained from Equation 10-5, with $n = 1$.

> theta := evalf(Pi-theta): e3a := theta = 376.99*t + 5.535: solve(e3a, t);

```
-.7117321539e-2
```

This time is still negative, but it is closer to the origin. We reuse the same sequence of commands, this time with theta a full revolution bigger than the original value.

> theta := arcsin(5/17.5): theta := evalf(2*Pi + theta): e3a := theta = 376.99*t + 5.535: solve(e3a, t);

```
.2753221595e-2 (2.75 ms)
```

We had to give theta its original value once more. That is the purpose of the command, > *theta := arcsin(5/17.5):*, which starts off the sequence. It might have been simpler to use the decimal value of theta and make up commands like > *solve (2*Pi + 0.28975 = 376.99*t + 5.535, t)*; instead of trying to reuse previous commands. Either way, the analytic solution provides a more accurate solution, yet we see that the graphical result is accurate to three figures.

Your turn. Find the smallest positive time when $v(t) = -50$ V if, in Equation 10-7, $A = 150$ V, $\omega = 100$ rad/s, and $\alpha = 4$ rad.

Answer: ______________________

Once again, we stress the graphical solution. It showed us the solution to this rather tricky problem in a highly visual way. In the analytic solution, we had to apply trigonometric identities and we had to redefine the value of theta by making it successively larger until the solution yielded a positive time.

Be Careful Not to Introduce Extra Solutions

Many authors suggest that when dealing with identities, you should work with each side of the equation but do not apply any operation to both sides at once. As a simple illustration, consider the trigonometric equation:

$$\sin(x) = -\sin(x) \qquad (10\text{-}8)$$

This equation does have a solution! What number is equal to its negative? The number can only be 0, since $0 = -0$ and no other number has this property. The solution must be $x = 0$, but is this the only solution? A graph provides an almost immediate answer (see Figure 10.5).

```
> plot( { sin(x),-sin(x) }, x = -4*Pi .. 4*Pi);
```

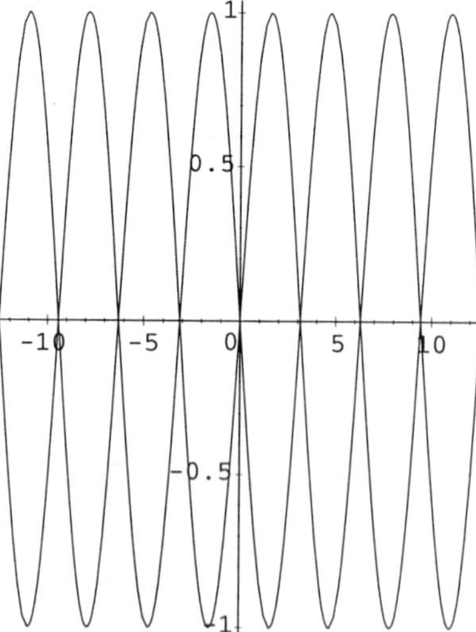

Figure 10.5 Solving Identities

The two curves are mirror images of one another. The x axis serves as the mirror. The graph shows that Eq. 10-8 has an infinite number of solutions, … $-3\pi, -2\pi, -\pi, 0, \pi, 2\pi, 3\pi$ …

What would have happened if you squared both sides? The equation becomes

$$\sin(x)^2 = \sin(x)^2 \qquad (10\text{-}9)$$

which is an identity! Normally squaring both sides of an equation introduces extra solutions, this time it introduced an infinite number of them.

Don't Throw Away Solutions

In the last section we saw that extra solutions can be introduced by squaring both sides of an equation. This time we will have to be careful or we will lose a solution. We choose the equation

$$\cos(\theta) = \left(\frac{3}{2} - \sin(\theta)\right)\cos(\theta) \qquad (10\text{-}10)$$

Canceling the common term, cos(θ), we have Maple solve the rest:

```
> solve(solve((3/2-sin(theta)=1));
                    1/6*Pi
```

But look what happens when we give Maple the original equation:

```
> eq := cos(theta)= (3/2-sin(theta))*cos(theta): solve(eq);
              1/2*Pi, 1/6*Pi, 5/6*Pi
```

Maple gives us more solutions. Where did they come from! We canceled the common term, but we are only allowed to do that if it is not zero. If $\cos(\theta) = 0$, then $\theta = 1/2*\text{Pi}$. This solution comes from the "discarded" cos(θ) term. The other solution comes from the solution to $\left(\frac{3}{2} - \sin(\theta)\right)$. This time, Maple decided to include the "$\pi - \theta$" solution as well.

Paper and Pencil Exercises

PP10–1

Solve the following trigonometric equations in the domain specified. (Some may have no solutions!) Use the basic definitions of the trigonometric functions to justify your answers.

(a) $\sin(x) = \cos(x) + 1, 0 \le x < 2\pi$ *Answer:* ____________

(b) $\sin(x) - 1 = \cos(x) + 1, 0 \le x < 2\pi$ *Answer:* ____________

(c) $\tan(x) = \cos(x), 0 \le x < 2\pi$ *Answer:* ____________

(d) $\sec(x) = \csc(x), 0 \le x < 2\pi$ *Answer:* ____________

(e) $\tan(x) = \sin(x), 0 \le x < 2\pi$ *Answer:* ____________

(f) $\sin(x) = \cos(x), 0 \le x < 360^\circ$ *Answer:* ____________

(g) $\tan(x) = \cos(x), 0 \le x < 360^\circ$ *Answer:* ____________

(h) $\cos(x) = -\cos(x), 0 \le x < 360^\circ$ *Answer:* ____________

PP10–2

Solve $\sin(\theta)(3/2 - \cos(\theta)) = \sin(\theta), 0 \le \theta < 2\pi$ *Answer:* ____________

PP10–3

Solve the following trigonometric equations in the domain specified:

(a) $\cos(A) = 1/2, 0 \le x < 2\pi$ *Answer:* ____________

(b) $2\cos(a) = -2, 0 \le x < 2\pi$ *Answer:* ____________

(c) $2\cos(a/2) = -2, 0 \le x < 2\pi$ *Answer:* ____________

(d) $2\sin(x) = \sqrt{3}, 0 \le x < \pi/2$ *Answer:* ____________

(e) $2\sin(x) = -\sqrt{3}, -2\pi \le x < 0$ *Answer:* ____________

(f) $\tan(2A) = -3/2, 0 \le x < 2\pi$ *Answer:* ____________

PP10–4

(a) Given $A\sin(x + y) = n$, find n if $A = 10$, $x = 30^\circ$, and $y = 45^\circ$.

Answer: ____________

(b) Given $\sin(a - b) = 3/4$, and $b = \pi/12$, find the smallest positive value of a that satisfies the equation.

Answer: ____________

Maple Lab

ML 10–1

The difficulty in solving trigonometric equations arises because there are usually an infinite number of solutions to a trigonometric equation.

Example 10-5. Consider the problem of finding all the solutions to the equation 2*cos(x) – 1 = 0:

> ex1 := 2*cos(x) - 1;

```
ex1 := 2 cos(x) - 1
```

We have a solution to the equation at every point where the graph of *ex*1 crosses the *x* axis, because we have $ex1 = 0$ at those places. The graph of the expression shows where it crosses the *x* axis (Figure 10.6):

> plot(ex1, x = -2*Pi .. 2*Pi);

The circles mark the solutions. You can see that if we continue the graph in either direction, we will find as many more solutions as we like. This raises an important point. We must know some other piece of information about the angle we are looking for if we want to limit the number of solutions.

We can let Maple find a solution. If we give Maple the equation as part of the *solve* command, we usually get only one of the infinite number of solutions.

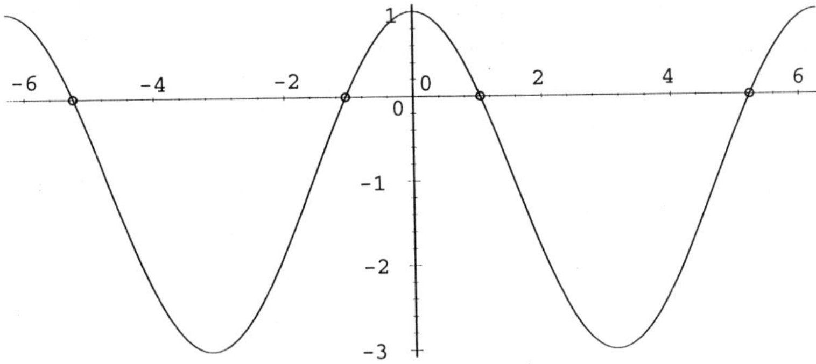

Figure 10.6 Graph of $y = 2*\cos(x)$ -1

> solve(ex1, x);

$$\frac{1}{3}\pi$$

Maple finds one solution. In fact, it is the smallest positive solution, at about $x = 1$. The other solutions can be found from this one, but we must go back to the fundamental definition of the cosine function if we want to figure out all the possible solutions. You know that cos(–*x*) = cos(*x*), so any solution you find immediately yields another solution, the negative of the first solution. Thus $-\frac{1}{3}\pi$ is also a solution to $y = 2*\cos(x) - 1$. To prove it, simply substitute this value in *ex*1 and verify that the expression evaluates to 0:

> ex1a := subs(x = -Pi/3, ex1);

$$ex1a := 2\cos(-\frac{1}{3}\pi) - 1$$

> simplify(ex1a);

$$0$$

We can write this as a *positive* angle if we want. This angle will still be less than 360°. The angle –*x* is the same as the angle 2*Pi –*x*, *x* in radians, or 360 – *x*, *x* in degrees. In our current problem, the current angle is –Pi/3, which we can express as +5*Pi/3, so 5*Pi/3 is also a solution. We verify that it is a solution as follows:

> ex1b := subs(x = 5*Pi/3, ex1);

$$ex1b := 2\cos\left(\frac{5}{3}\pi\right) - 1$$

> simplify(ex1b);

$$0$$

Now we can find all solutions. Adding a full rotation, or any number of full rotations, to our two already found solutions should produce a solution. Thus, the total solution set is:

$$\left\{\frac{1}{3}\pi + 2n\pi,\ \frac{5}{3}\pi + 2n\pi\right\},\quad n = 0,\ \pm 1,\ \pm 2,\ \pm 3,\ \text{etc}$$

where *n* is any integer. You can check any of these solutions. For example, take $n = 6$.

> ex1c := subs(x = 1/3*Pi + 2*6*Pi, ex1);

$$ex1c := 2\cos\left(\frac{37}{3}\pi\right) - 1$$

> simplify(ex1c);

$$0$$

Your turn.

(a) Find all solutions to 3 cos(x) – 2 = 0. State the exact result, not the decimal approximation.

Answer: ______________________________

(b) Find all solutions to 17 sin(x) – 18 = 0

Answer: ______________________________

(c) Plot the left hand side of the equation in part (b). Does the plot confirm your answer, or do you need to modify it?

Answer: ______________________________

ML10–2

Solve the equation 5*(1 + sin(*x*)) = 2*sin(*x*) + 3.

The solution must be an angle between 0 and 2*Pi:

> ex2a := 5*(1 + sin(x)) = 2*sin(x) + 3;

First, you should solve for sin(*x*):

> ex2b := solve(ex2, sin(x));

This says that sin(*x*) = –2/3, so we make this our next equation:

> ex2c := sin(x) = -2/3;

We could solve this simply by calculating arcsin(–2/3), or we could use the *solve* command:

> ex2d := solve(ex2c, x);

We get the same result! Maple cannot reduce the solution any further and still be completely exact! We use *evalf* to get a numerical solution:

> ex2e := evalf(ex2d);

An angle of –x is the same as the angle 2*Pi + x, so another solution is:

> ex2f := evalf(2*Pi + ex2e);

This angle is between 0 and 2*Pi, so it is one of the solutions we are looking for. The other solution is found by observing that:

```
sin(x)  =  sin(Pi  -  x)
```

> ex2g := evalf(Pi-ex2e);

Finally, we substitute our solutions into the original equation to verify:

> **evalf(subs(x = ex2f, ex2a));**

> **evalf(subs (x = ex2g, ex2a));**

Both solutions satisfy the original equation and are between 0 and 2*Pi. Thus, they are the desired solutions.

Your turn. Follow the preceding steps to find the solution to the equation

> **eq := -5*(1 - sin(x)) + 6 = 2*sin(x) + 3;**

The solution must be an angle between $-\pi$ and π.

Answer: ______________________________

ML10-3

Solve the equation, $2\cos(x)^2 - \sin(x) - 1 = 0$ ($0 <= x <= 2$*Pi):

> **ex3 := 2*cos(x)^2-sin(x)-1 = 0;**

The usual method requires that we substitute for cos(x)^2 to get an equation in terms of sin(x) only. Let's let Maple loose on the equation as it stands:

> **ex3a := solve(ex3, x);**

Not bad, the only trouble is that Maple did not find *all* solutions, and it reported one negative solution. Instead of putting the solutions in the range from 0 to 2*Pi, it puts the solutions it finds in the range –Pi .. +Pi.

Note. It is always wise to graph the equation to attempt to find solutions. But Maple cannot plot equations, just expressions! The solution is to graph the left-hand side of the equation. The places where this curve crosses the x axis are the solutions for which we are looking (Figure 10.7).

> **plot(lhs(ex3) , x = 0 .. 2*Pi);**

To more closely investigate the solutions around $x = 5$, we expand the plot in the range 4.5 .. 4.9:

> **plot(lhs(ex3), x = 4.5 .. 4.9);**

It begins to look as if the curve just touched the x axis. Since it doesn't cross the axis, there is only one solution: the point where it touches. This point is at $x = 4.71$. We can verify this solution as well:

> **evalf(subs(x = 4.71, ex3));**

The agreement is not exact, so we could adjust the number 4.71 slightly.

> evalf(subs(x = 4.712, ex3));

Thus, $x = 4.712$ gives an even better match. You might want to apply the ideas expressed in Example 10-2 to find the exact value.

In addition to solving the equations analytically, provide a graphical solution and check each solution by back-substitution:

Find all solutions in the range $0 \le x \le 2\pi$ to equation *eq*3 by following the method outlined above.

> eq3 := 2*sin(x)^2-cos(x)-1=0

Answer: ____

ML10–4

Solve 2*sin(*x*) + 1 – 0, 0 <= *x* <= 2*Pi *Answer:* ____________________

ML10–5

Solve 2*cos(*x*)^2 – 2*cos(2**x*) – 1 = 0, 0 <= *x* <= 2*Pi

Answer: ____

ML10–6

Solve 4*tan(*x*) – sec(*x*)^2 = 0, 0 <= *x* <= 2*Pi *Answer:* ____________________

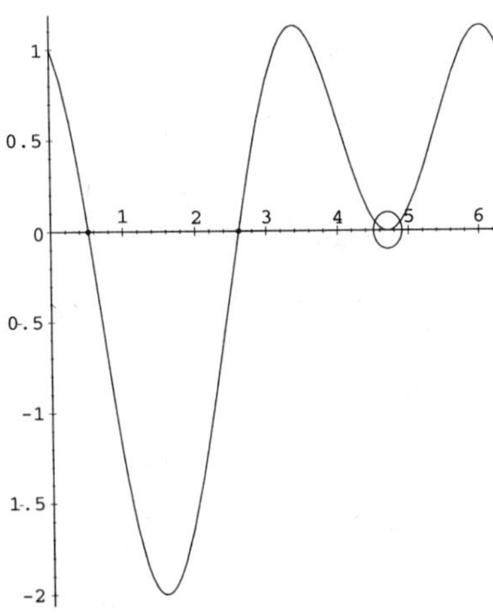

Figure 10.7 The Solutions to 2*cos(x)^2 - sin(x) - 1 = 0

ML10-7

Solve cos(3*x)*cos(x) – sin(3*x)*sin(x) = 0, 0 <= x <= 2*Pi

Answer: ____________________

ML10-8

State which of the following are identities (answer true or false):

(a) sin(x)*(csc(x) – sin(x)) = cos(x)^2 *Answer:* __________

(b) cos(y)^2 – sin(y)^2 = (1 – tan(y)^2)/(1+tan(y)^2) *Answer:* __________

(c) (1+cos(2*x))/cos(x)^2 = 1 *Answer:* __________

(d) sin(x/2)*cos(x/2) = (sin(x))/2 *Answer:* __________

ML10-9

You are designing a robot arm as part of an electronics project. You develop an equation for the critical angle between two sections of the robot arm:

eq9 := cos(theta) + 1.3 * cos(2*theta) = 0;

(a) Plot the left-hand side of eq9 to determine a solution in the range:

0 <= theta <= Pi

Sketch the plot here:

Answers: The solution, as read from the plot, is: __________ rad.

This solution, converted to degrees, is: __________ deg

(b) Solve the equation using the *solve* command.

Answers: The solution given by *solve* is: __________ rad.

The solution given by *solve* is: __________ deg.

Index

A

B

C